Sustainable Civil Infrastructures

Sustainable Civil Infrastructures (SUCI) is a series of peer-reviewed books and proceedings based on the best studies on emerging research from all fields related to sustainable infrastructures and aiming at improving our well-being and day-to-day lives. The infrastructures we are building today will shape our lives tomorrow. The complex and diverse nature of the impacts due to weather extremes on transportation and civil infrastructures can be seen in our roadways, bridges, and buildings. Extreme summer temperatures, droughts, flash floods, and rising numbers of freeze-thaw cycles pose challenges for civil infrastructure and can endanger public safety. We constantly hear how civil infrastructures need constant attention, preservation, and upgrading. Such improvements and developments would obviously benefit from our desired book series that provide sustainable engineering materials and designs. The economic impact is huge and much research has been conducted worldwide. The future holds many opportunities, not only for researchers in a given country, but also for the worldwide field engineers who apply and implement these technologies. We believe that no approach can succeed if it does not unite the efforts of various engineering disciplines from all over the world under one umbrella to offer a beacon of modern solutions to the global infrastructure. Experts from the various engineering disciplines around the globe will participate in this series, including: Geotechnical, Geological, Geoscience, Petroleum, Structural, Transportation, Bridge, Infrastructure, Energy, Architectural, Chemical and Materials, and other related Engineering disciplines.

SUCI series is now indexed in SCOPUS and EI Compendex.

More information about this series at http://www.springer.com/series/15140

S. Sonny Kim · Arif Ali Baig Moghal ·
Jia-liang Yao
Editors

Advances in Urban Geotechnical Engineering

Proceedings of the 6th GeoChina International Conference on Civil & Transportation Infrastructures: From Engineering to Smart & Green Life Cycle Solutions – Nanchang, China, 2021

Springer

Editors
S. Sonny Kim
School of Environmental, Civil, Agricultural, and Mechanical Engineering
University of Georgia
Athens, USA

Arif Ali Baig Moghal
Department of Civil Engineering
National Institute of Technology
Warangal, India

Jia-liang Yao
School of Traffic and Transportation Engineering
Changsha University of Science and Technology
Changsha, China

ISSN 2366-3405 ISSN 2366-3413 (electronic)
Sustainable Civil Infrastructures
ISBN 978-3-030-80151-9 ISBN 978-3-030-80152-6 (eBook)
https://doi.org/10.1007/978-3-030-80152-6

This Springer imprint is published by the registered company Springer Nature Switzerland AG
The registered company address is: Gewerbestrasse 11, 6330 Cham, Switzerland

Introduction

This volume contains eight papers that were accepted and presented at the GeoChina 2021 International Conference on Civil and Transportation Infrastructures: From Engineering to Smart and Green Life Cycle Solutions, held in NanChang, China, September 18 to 19, 2021. It contains research data, discussions, and conclusions focusing on a compilation of studies regarding transportation geotechnics, geomechanics, rock mechanics, and geosynthetics-reinforced soils. Topics include issues related to the advancements on new technologies and materials as solutions for the challenges in the geotechnical engineering. This information should shed the lights on new findings obtained from laboratory tests, field application, and numerical simulation, which can strongly promote the understandings on the soils earth behaviors. It is anticipated that this volume will present the advancements on the numerical simulation method in the geotechnical engineering to support a more resilient environment for infrastructure users.

Contents

About the Editors

Dr. Prof. S. Sonny Kim is an associate professor of Civil Engineering at the University of Georgia. He received his M.S. from Georgia Tech in 2000 and Ph.D from Texas A&M University in 2004. His primary research has focused on transportation geotechnics, geosynthetics-reinforced pavement foundation, and performance evaluation of aggregate/subgrade layers using nondestructive test (NDT). Before joining the academia, he has been an integral contributor to several nationwide highway/airport pavement design and construction projects.

Dr. Kim is a fellow of the ASCE and was also named Distinguished Faculty Fellow by College of Engineering from the University of Georgia. Dr. Kim is an active member of American Society of Civil Engineers (ASCE) G-I Highway Pavement Committee, ASCE T&DI Committee, and Transportation Research Board (TRB) AKP 20 and AKM 80 Committee. He is a licensed professional engineer and has written over 100 refereed articles and reports.

Arif Ali Baig Moghal is an active consultant, researcher, and academician. He obtained his Ph.D. from Indian Institute of Science, Bangalore, India. He is currently an associate professor in the Department of Civil Engineering at National Institute of Technology, Warangal, India, and had worked earlier at King Saud University, Riyadh, Saudi Arabia. He was responsible for establishing state-of-the-art Geoenvironmental Engineering Laboratory in the Department of Civil Engineering at King Saud University, Riyadh, during his tenure there. He is an active member of American Society of Civil Engineers (ASCE) and Indian Geotechnical Society (IGS). He has authored, co-authored over one hundred articles in peer-reviewed journals, book chapters, and conference proceedings. (ORCID: 0000-0001-8623-7102).

Prof. Jia-liang Yao is a deputy director of Institute of transportation economy and a professor of Materials Department of School of transportation engineering of Changsha University of technology. He is a professional registered road supervision engineer, and his research direction is pavement engineering and pavement materials. He completed his undergraduate and graduate studies in Hunan University,

China. He has written, co-authored, and edited 45 journal papers, 12 monographs and textbooks (author / co-author / editor), and 20 conference papers. He presided over and participated in 26 provincial and ministerial level projects and served as a technical consultant for several expressway projects.

Research on Soil Characteristics and Controlling Technology of Differential Deformation of Expressway Widening Subgrade in Seasonal Frozen Region

Shengchuan Liu[1(✉)], Liming Yu[2], and Guangyuan Jiang[3]

[1] Ministry of Transport Research Institute of Highways, Beijing 100088, China
[2] Xinjiang Beixin Sifang Engineering Testing & Consulting Co, Ltd, Wurumuqi 830000, China
[3] Xinjiang Beixin Road and Bridge Group Co., Ltd, Wurumuqi 830001, China

Abstract. This paper is aiming at soil characteristics and controlling technology of differential deformation of expressway widening subgrade in seasonal frozen region. The deformation of widened subgrade is the key factor affecting the service life and driving comfort of expressway after widening. Therefore, it is of special theoretical and practical significance to carry out research on influence of expressway widening on subgrade deformation in seasonal frozen region, which effectively control the overall stability and uneven settlement of the new and old subgrade, reduce the damage of the upper pavement structure layer caused by this, and ensure the quality of widening expressway. Through a series of triaxial tests, the relationship between the strength and the stress-strain of the soil is studied. And the influences of the number of freezing and thawing, the test confining pressure, the cooling temperature and the freezing mode on the stress-strain relationship and the strength characteristics are analyzed, and the stress-strain relationship of the subgrade filling soil is described. The finite element method had been carried out to simulate the deformation of widening subgrade. And the influence of each condition of Highway Widening on the subgrade deformation is analyzed, especially the deformation features of new and old subgrade. According to the results, frost heave deformation control standard of cohesive soil subgrade had been proposed. At last, controlling technology of differential settlement and construction suggestions of expressway widening subgrade had been put forward in seasonal frozen region.

Keywords: Widening subgrade · Control standard · Differential deformation · Triaxial test · Seasonal frozen region

1 Introduction

The deformation of widened subgrade is the key factor affecting the service life and driving comfort of expressway after widening. But in the road expansion process, the problem of the subgrade widened, especially widen embankment deformation properties and stability has become increasingly prominent. Highway widen project will be problem

S. S. Kim et al. (Eds.): GeoChina 2021, SUCI, pp. 1–15, 2021.
https://doi.org/10.1007/978-3-030-80152-6_1

that our highway construction must face and urgently settle at the beginning of this century. Therefore, it is of special theoretical and practical significance to carry out research on influence of expressway widening on subgrade deformation in seasonal frozen region, which effectively control the overall stability and uneven settlement of the new and old subgrade, reduce the damage of the upper pavement structure layer caused by this, and ensure the quality of widening expressway. Ling Jianming and other researchers [1–6] have carry out systematic study on the deformation behavior and stabilized technology of the new and old embankment employing the site tests, numerical simulation and theoretical analysis, etc. methods. Consider widening the road pavement to meet functional requirements, differential settlement control indicators have been proposed in some papers [7–10] based on the brief analysis and specification.

This paper, the relationship between the strength and the stress-strain of the soil is studied through a series of triaxial tests. And the influences of the number of freezing and thawing, the test confining pressure, the cooling temperature and the freezing mode on the stress-strain relationship and the strength characteristics are analyzed, and the stress-strain relationship of the subgrade filling soil is described. The finite element method had been carried out to simulate the deformation of widening subgrade. And the influence of cach condition of Highway Widening on the subgrade deformation is analyzed, especially the deformation features of new and old subgrade. According to the results, frost heave deformation control standard of cohesive soil subgrade had been proposed. At last, controlling technology of differential settlement and construction suggestions of expressway widening subgrade had been put forward in seasonal frozen region.

2 Study on Static Characteristics of Subgrade Filling Soil Under Freeze-Thaw Condition

2.1 Method of Freeze-Thaw Cycles and Test Materials

The determination of the freezing mode affects the mechanical properties of the sample after the freeze-thaw cycles. This test uses a triaxial freezing method with the same freezing rate in a closed system. The freeze-thaw cycle process is carried out in the freeze-thaw test chamber, as shown in Fig. 1. In order to ensure that the moisture content of the sample does not change during the freezing and thawing process, this test uses a plastic wrap to keep it moisturized so that the freezing and thawing process is close to a closed condition and can ensure that the moisture content changes little.

Through indoor geotechnical tests, the mechanical parameters of the soil samples are obtained. The test results show that the soil samples have a better gradation, with a plasticity index < 10, and are silty soils.

2.2 Static Triaxial Test Analysis of Plain Soil and Subgrade Filling Soil Before Freezing and Thawing

It can be seen that the plain soil sample is in the elastic deformation stage within 0.6% of the strain, and unrecoverable plastic deformation appears when the strain exceeds

0.6% through static triaxial test (Fig. 1). Cracks appears inside and with the continuous development of the cracks, the sample reaches the failure strength at 4.5% when the sample is destroyed. The deviator stress at failure is 262 kPa.

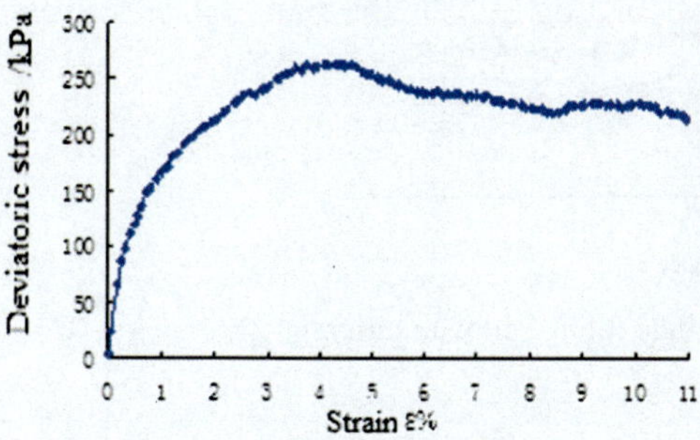

Fig. 1. Stress-strain relationship of plain soil after 0-cycle

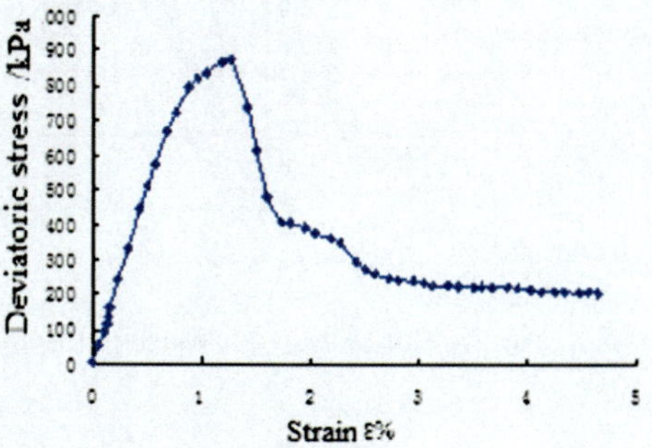

Fig. 2. Stress-strain relationship of subgrade filling soil after 0-cycle

It can be seen that the subgrade filling soil sample is in the elastic deformation stage within 1.1% of the strain, and unrecoverable plastic deformation appears when the strain exceeds 1.1% through static triaxial test (Fig. 2). The sample suddenly fails when the strain reaches 1.4%, and the ultimate stress during failure is 876 kPa. The stress decreases sharply with the increase of strain, which shows the typical failure properties of subgrade filling soil. Comparing the ultimate stress of two soil samples, the ultimate stress of the subgrade filling soil is 3 times the strength of the plain soil, and the elastic deformation stage of the improved soil is more obvious.

2.3 Influence of Freeze-Thaw Cycles on Stress-Strain Relationship of Subgrade Filling Soil

The stress-strain relationship of subgrade filling soil under the same freeze-thaw temperature and different cycles is shown in Fig. 3, Fig. 4, Fig. 5, Fig. 6 and Fig. 7.

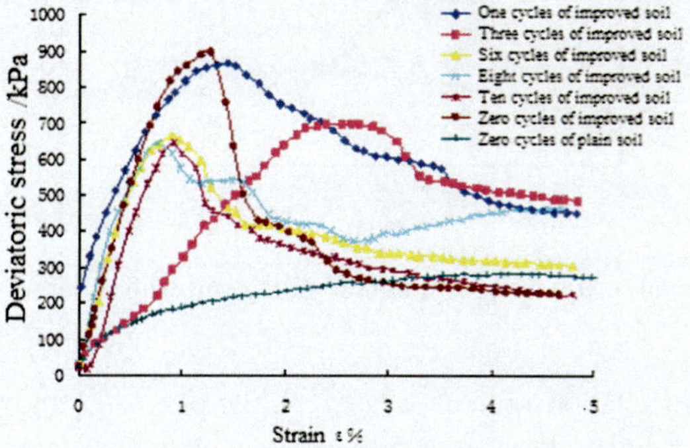

Fig. 3. Stress strain relationship of subgrade filling soil with different cycles at −2 °C

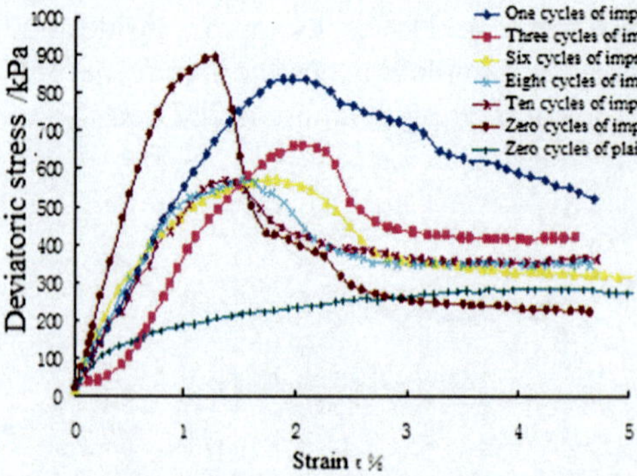

Fig. 4. Stress strain relationship of subgrade filling soil with different cycles at −5 °C

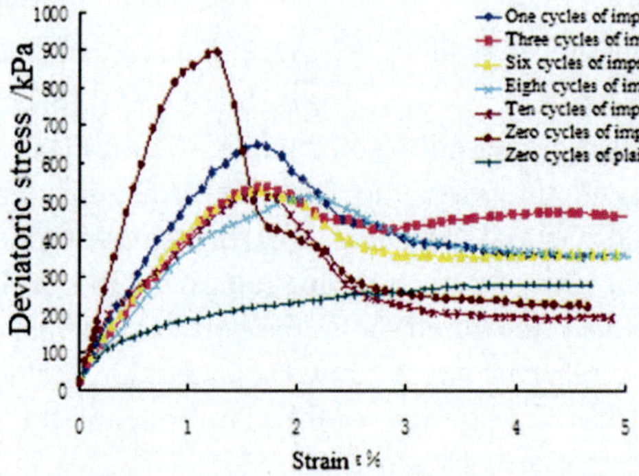

Fig. 5. Stress strain relationship of subgrade filling soil with different cycles at −10 °C

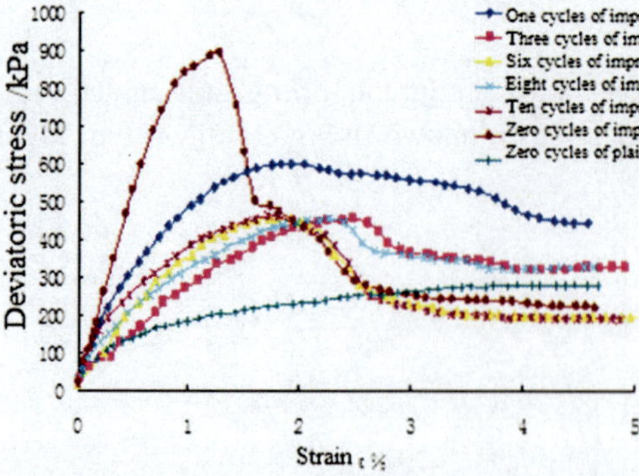

Fig. 6. Stress strain relationship of subgrade filling soil with different cycles at −15 °C

The stress-strain relationship curve between the plain soil and the subgrade filling soil is added in the figure without freeze-thaw cycle, which is compared with the stress-strain relationship of subgrade filling soil after cycling. It can be seen that when the confining pressure is 20 kPa, the damage of one freeze-thaw cycle on subgrade filling soil is small, and the limit stress of subgrade filling soil changes after three cycles when

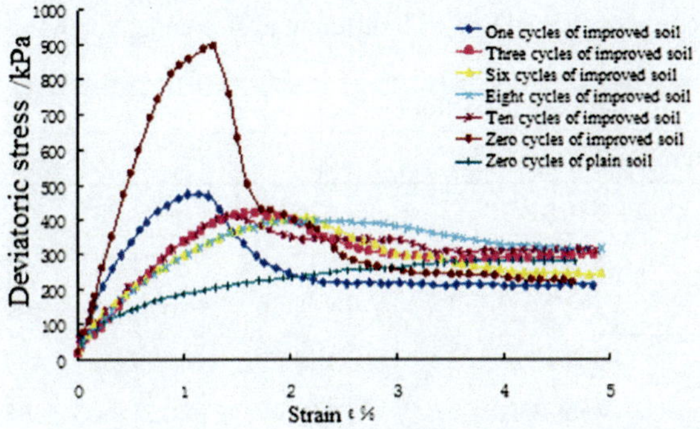

Fig. 7. Stress strain relationship of subgrade filling soil with different cycles at −20 °C

the freezing temperature is −2 °C and −5 °C. After six cycles, the curve of the sample is basically stable, and the peak value of deviatoric stress changes little. When the freezing temperature is −10 °C and −15 °C, the results show that the peak strength of the first cycle has a significant change compared with that of without cycle, which decreases by about 1/3. The strength of the third cycle continues to decrease compared with that of the first cycle, and the ultimate deviator stress is basically stable after more than three cycles.

The data image is similar and there is no big change any more. It indicates that the effect of freeze-thaw on the subgrade filling soil will not change after three cycles in this temperature range.

Under the freezing condition of −20 °C, it can be seen from the figure that the ultimate deviatoric stress of the subgrade filling soil is reduced by 1/2 after one freeze-thaw cycle, and the failure stress of the third cycle is less than that of the first cycle, and the data changes little after more than three cycles, which indicates that the subgrade filling soil of subgrade filling basically enters a stable state after one cycle.

2.4 Static Strength Characteristics of Plain Soil and Subgrade Filling Soil

According to the test data, the failure stress of triaxial test under different freeze-thaw temperature is shown in Table 1.

The relationship between static strength of subgrade filling soil and freeze-thaw times is shown in Fig. 8.

It can be seen from the figure that the damage degree of one cycle freeze-thaw on the subgrade filling soil is small under one cycle of −2 °C and −5 °C. In addition to the larger static strength of the subgrade filling soil under negative temperature of −2 °C and −5 °C, the static strength of the subgrade filling soil decreases rapidly between −5 °C and −10 °C. After more than 6 cycles and up to 10 cycles, the change of limit stress at each temperature is similar. This is consistent with the curves of unconfined compressive strength at different temperatures.

Table 1. Failure stress of triaxial test with different cycles under confining pressure of 20 kPa

Peak stress	Freeze thaw times				
Freezing temperature	1	3	6	8	10
−2 °C	846 kPa	676 kPa	640 kPa	622 kPa	637 kPa
−5 °C	819 kPa	643 kPa	551 kPa	548 kPa	550 kPa
−10 °C	631 kPa	516 kPa	507 kPa	499 kPa	494 kPa
−15 °C	581 kPa	437 kPa	435 kPa	437 kPa	445 kPa
−20 °C	453 kPa	418 kPa	402 kPa	395 kPa	406 kPa

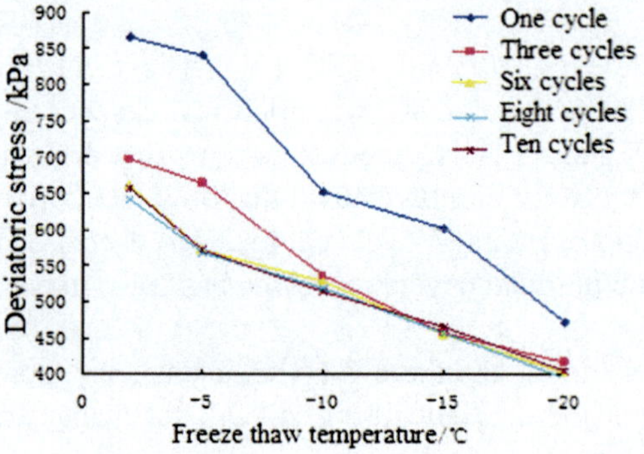

Fig. 8. Relationship between static strength of filling soil and freeze-thaw times

3 Numerical Simulation Analysis of Differential Settlement of Old and New Subgrade

The finite element method had been carried out to simulate numerical values in different kinds of conditions, such as, the widening ways, excavation and filling of steps, filling height, filling rate, using geogrid. The deformation laws of widening subgrade had been found under different conditions and influential factors.

In the calculation model, the embankment section is analyzed by two-dimensional plane finite element method, and the calculated width and depth of foundation are 140 m and 25 m respectively. Mohr-Coulomb model is used for foundation soil, subgrade soil and gravel cushion, and elastic-plastic model is used for pile and geogrid. Considering the consolidation of the soil below the groundwater level, and the Biot model is selected as the consolidation model. The equilibrium initial stress field is established first, and then the embankment is loaded step by step. After the completion of the project, 20 kPa uniform load was applied to simulate the road surface and vehicle load. The road surface and vehicle load are simulated by 20 kPa uniform load. In the initial stress balance, the uniform load is applied to the surface of the old subgrade to simulate the dynamic compaction process. The length of uniform load is 1.5 m and the interval is 5 m. When $t = 0.042$ s, the load reaches the maximum, and the total contact time between the load and the surface is 0.084 s (Table 2 and Table 3).

Table 2. Material calculation parameters

	Elastic modulus (MPa)	Cohesion (kPa)	Poisson's ratio	Friction angle (°)	Soil gravity (KN/m^3)
Filling soil	23	30	0.3	20	19
Gravel cushion	50	20	0.3	30	20
Gravel pile	120		0.2		22

Table 3. Soil parameters

Soil layer	Depth (m)	Void ratio	Permeability coefficient (m/d)	Young's modulus (MPa)	Bulk density	Poisson's ratio
Clayey sand	1	0.829	1.2e−2	5	17.2	0.3
Silty clay	10	0.79	7.3e−4	8	19	0.3
Clay	25	0.6	6.1e−5	16.2	20.4	0.3

In the simulation of pile, the analysis step is set after stress balance to simulate the loading of pile by increasing the extra load of pile compared with soil because the gravity of gravel pile is different from that of soil. Truss model is used to simulate the geogrid with a width of 8 m laid on the top three steps.

Through the steps of establishing the geometric model, determining the material model, selecting the calculation parameters and simulating the construction loading process, the deformation characteristics of the widened subgrade can be seen intuitively through the finite element calculation, as shown in Fig. 9 and Fig. 10. The deformation values of the old subgrade, the new subgrade and the joint of the old subgrade and the new subgrade can be obtained, including the lateral displacement and vertical settlement, as shown in Fig. 11, Fig. 12, Fig. 13 and Fig. 14.

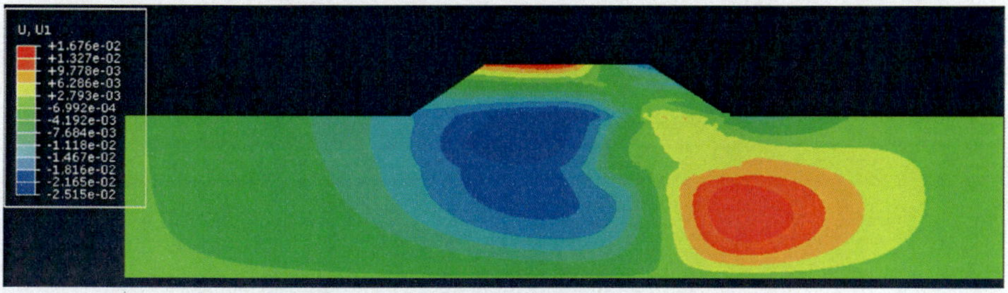

Fig. 9. Coloring diagram of lateral displacement in 15 years after construction

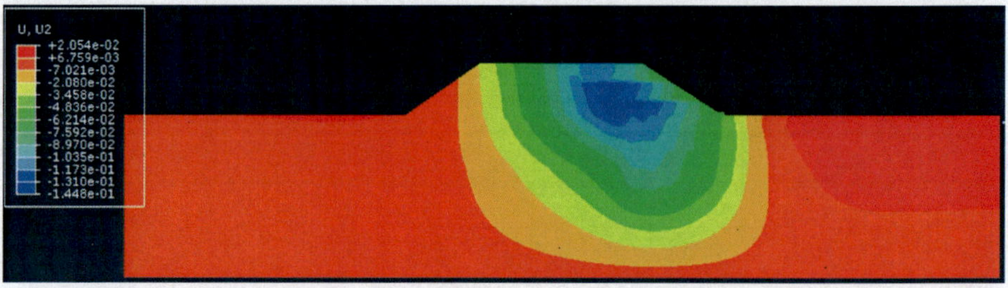

Fig. 10. Coloring diagram of vertical settlement in 15 years after construction

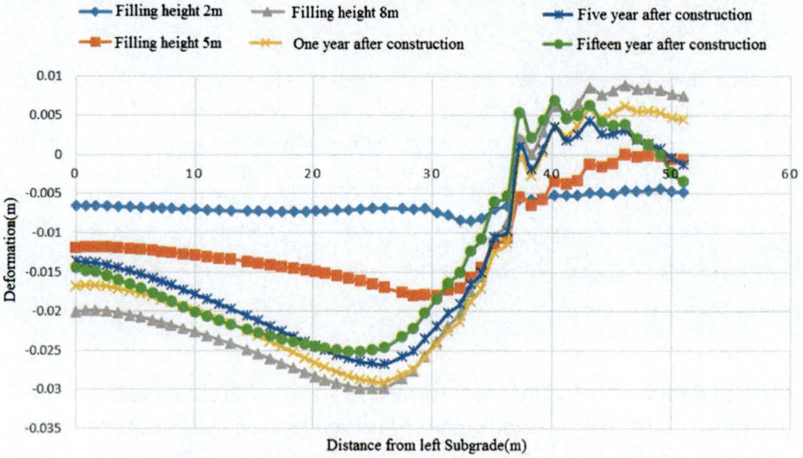

Fig. 11. Lateral displacement curve of ground surface

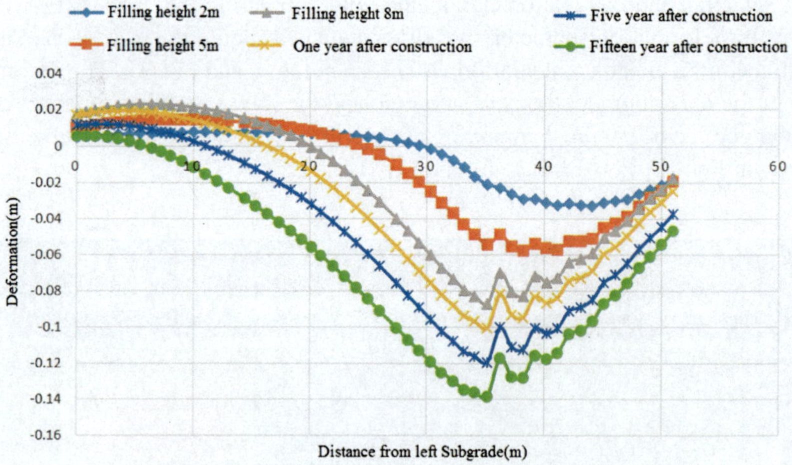

Fig. 12. Vertical displacement curve of ground surface

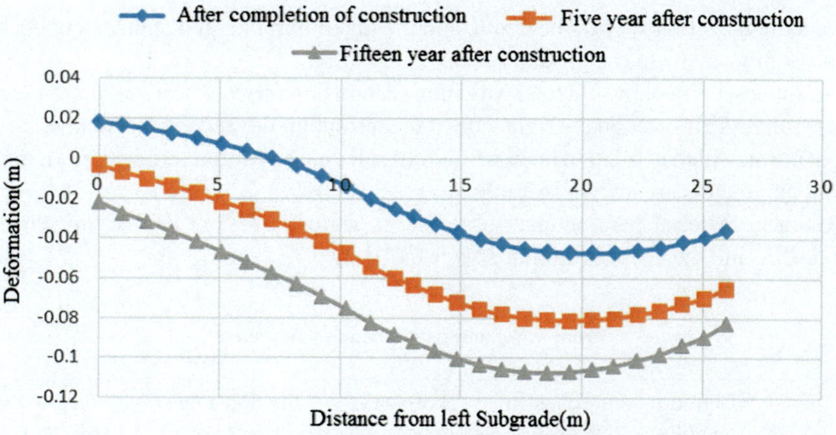

Fig. 13. Vertical displacement curve of road surface

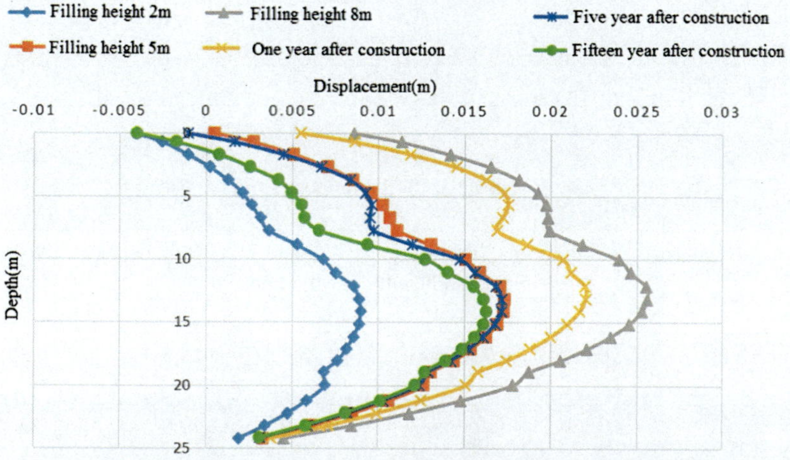

Fig. 14. Lateral displacement curve of slope toe of new subgrade

According to the development trend of the vertical displacement curve, the deformation first increases and then decreases with the increase of the distance from the left shoulder. The maximum surface settlement after construction is 8.7 cm, and the maximum settlement in 15 years is 13.8 cm, which are all less than 20 cm. There is an obvious mutation in the middle of the curve, which indicates that there is a differential settlement between the old and the new subgrade after 15 years of construction, and the value is less than 3 cm. The maximum settlement of road surface is 4.67 cm after construction and 10.7 cm in 15 years after construction, and there is no obvious differential settlement at the joint of new and old subgrade. The horizontal displacement of this section is relatively small, and the maximum lateral displacement of the ground surface and the foot of embankment are 2.51 cm and 1.63 cm respectively 15 years after construction. The displacement of the old subgrade is to the left, and the displacement of the new subgrade is to the right. The horizontal displacement of the new subgrade slope toe increases first and then decreases with the depth, and the maximum lateral displacement occurs at about 13 m away from the ground surface.

Another section of the old road widening section has a serious soft foundation section. A new subgrade is built on the right side of the old subgrade, and the embankment filling height is 8m, which is filled in layers. Material calculation parameters are shown in Table 4. Soil parameters are shown in Table 5.

Geogrid material parameters: the tensile modulus is 3.87 * 10^4 kN/m, Poisson's ratio is 0.3, and the cross-sectional area is 0.002 m^2.

Table 4. Material calculation parameters

	Elastic modulus (MPa)	Cohesion (kPa)	Poisson's ratio	Friction angle (°)	Soil gravity (KN/m^3)
Filling soil	23	30	0.3	20	19
Gravel cushion	50	20	0.3	30	20
Gravel pile	120		0.2		22

Table 5. Soil parameters

Soil layer	Depth (m)	Void ratio	Permeability coefficient (m/d)	Young's modulus (MPa)	Bulk density	Poisson's ratio
Clayey sand-1	1	0.63	5.5e−3	15	18.2	0.3
Clayey sand-2	2	0.391	5.9e−3	31.9	20.6	0.3
Silty clay-3	10	0.606	9e−4	9.5	20.6	0.3
Clay-4	15	0.680	9.7e−4	19	20.4	0.3
Clay-5	18	0.800	9.4e−5	18	19.7	0.3
Silty clay-6	25	0.602	8.6e−4	14.4	20.5	0.3

The deformation characteristics of the widened subgrade can be seen intuitively through the finite element calculation, as shown in Fig. 15 and Fig. 16.

The deformation values of the old subgrade, the new subgrade and the joint of the old subgrade and the new subgrade can be obtained, including the lateral displacement and vertical settlement, as shown in Fig. 17, Fig. 18, Fig. 19 and Fig. 20.

It can be seen from the vertical displacement curve, the deformation first increases and then decreases with the increase of the distance from the left shoulder. The maximum

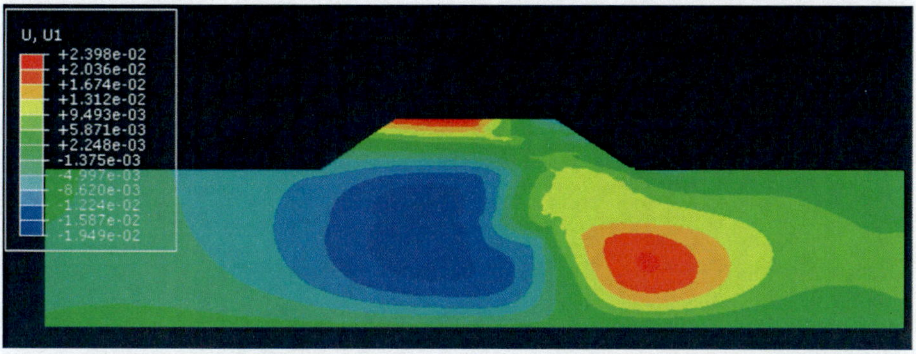

Fig. 15. Coloring diagram of lateral displacement in 15 years after construction

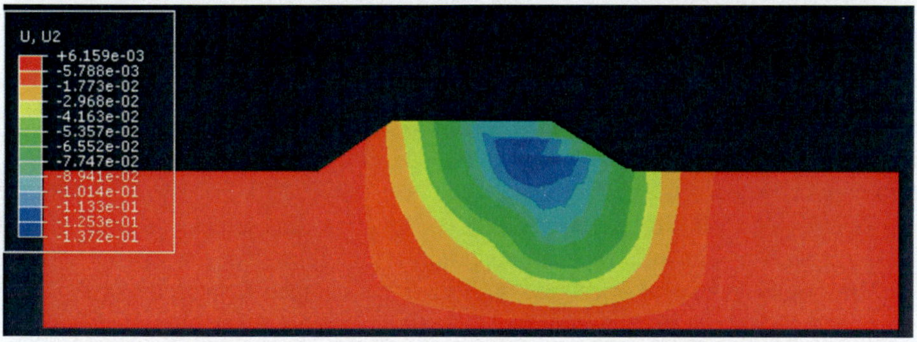

Fig. 16. Coloring diagram of vertical settlement in 15 years after construction

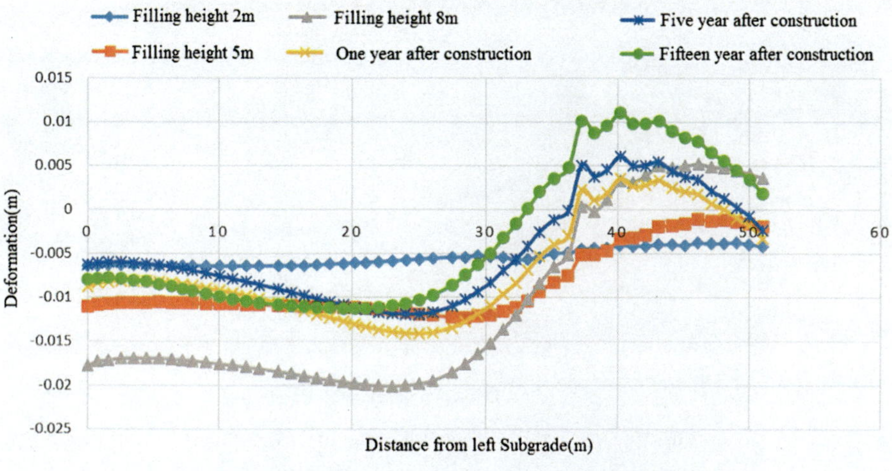

Fig. 17. Lateral displacement curve of ground surface

vertical displacement of the surface is 8.455 cm and the uneven settlement is 10.75 cm, and the maximum settlement is 12.503 cm and the uneven settlement is 12.558 cm in

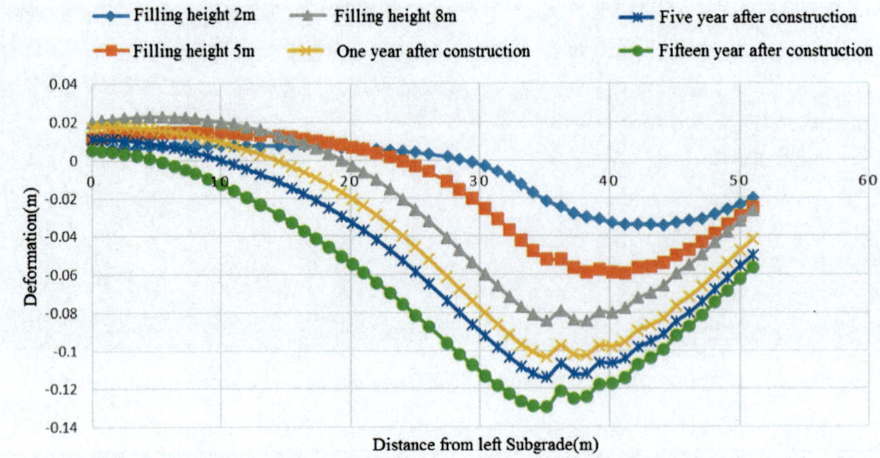

Fig. 18. Vertical displacement curve of ground surface

15 years after construction. After the completion of construction, the settlement gradually tends to be stable. The settlement of road surface is 4.503 cm and the differential settlement is 4.664 cm. The maximum settlement was 10.043 cm and the differential settlement was 7.701 cm in 15 years after construction. The settlement curve of road surface is smooth without obvious discontinuity which shows that the settlement of the road surface is within the specified range after the completion of construction. And there is no obvious difference settlement at the joint of the new and old subgrade which indicates that the overlapping effect is good.

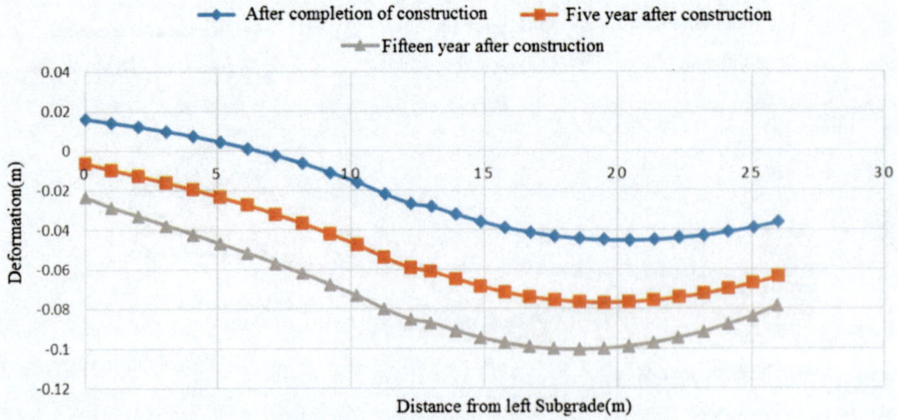

Fig. 19. Vertical displacement curve of road surface

The surface horizontal displacement and the horizontal displacement under the slope toe are relatively small, and the maximum displacement is within 2 cm. The maximum horizontal displacement of the ground surface is 1.998 cm after the completion of the construction, and the lateral displacement direction of the foundation under the new and

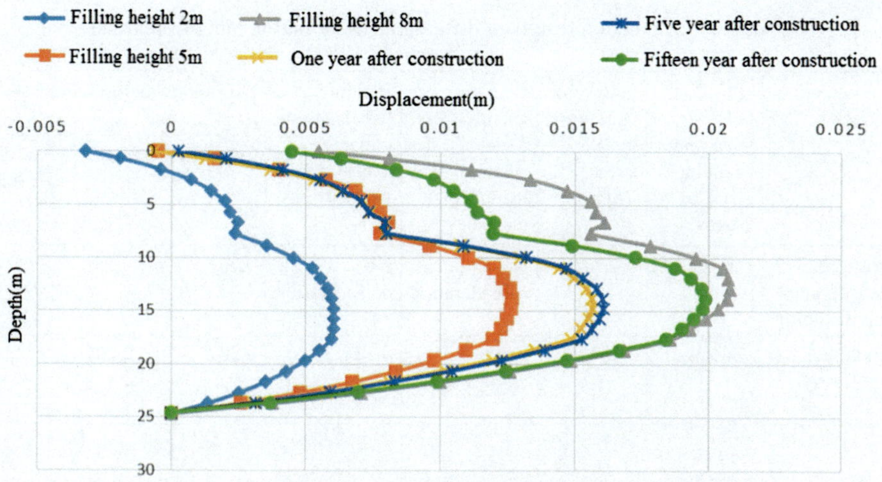

Fig. 20. Lateral displacement curve of slope toe of new subgrade

old subgrade is opposite. The maximum displacement under the slope toe is 2.081 cm after the completion of the construction, which gradually decreases and tends to be stable after construction and there is a maximum lateral displacement within 10–15 m underground. Overall, the horizontal displacement is small and the maximum horizontal displacement of soil is 2.398 cm, which has little impact on the normal use of subgrade.

4 Differential Deformation Control Standard of New and Old Subgrade

The calculation results of differential deformation rate of subgrade based on design service life (15 years) are shown in Table 6. From the calculation results, the differential deformation rate of subgrade should be less than 2.61%. Therefore, considering the standard driving load, the differential deformation rate of subgrade should be less than 3.07%, so as to ensure the service safety within the design life. Considering the overload load, the subgrade's differential deformation rate should be less than 3.07% and the differential deformation rate must be less than 2.27%.

According to the fatigue equation of asphalt surface layer and semi-rigid material obtained from the split fatigue experiment of domestic asphalt and semi-rigid material, the service life of cohesive soil subgrade in the seasonal frozen area can be calculated from the different deformation rate of subgrade. The relationship between differential deformation rate of subgrade top and service life is shown in Table 7.

From the calculation results, only considering the frost heave factor, when the differential deformation rate of subgrade is 1%, the service life of highway can reach 86 years under the standard driving load and 46 years under the overload effect, while when the differential deformation rate of subgrade is 4%, the service life of both loads is less than 1 year. From the relationship between differential deformation rate and service life, it can be seen that when the deformation rate is more than 3%, the relationship between the two is linear decreasing; when the deformation rate is more than 3%, the service life

Table 6. Calculation results of differential deformation rate of subgrade

Computational model	Position of maximum bending tensile stress	Allowable bending tensile stress /MPa	Bending tensile stress under standard axle load /Mpa	Bending tensile stress under overload axle load /Mpa	Differential deformation rate /%	
					Standard axle load	Overload axle load
Positive deflection of middle uplift	Upper layer	0.424	No consideration	No consideration	0.261	
Shoulder uplift reverse deflection	Subbase	0.284	0.096	0.144	0.307	0.227

Table 7. Life prediction of highway under different deformation rates

Differential deformation rate	Maximum bending tensile stress /MPa	Splitting strength/MPa	Number of fatigue actions / ten thousand times		Service life /years	
			Standard axle load	Overload axle load	Standard axle load	Overload axle load
1%	0.064	0.6	1.43E + 06	3.32E + 04	86.1	46.7
2%	0.121	0.6	1.81E + 04	1.03E + 03	40.5	13.5
3%	0.223	0.6	71.85	9.62	1.8	0.3
4%	0.278	0.6	7.34	1.30	0.2	0.0

changes little to zero. Considering the allowable bending tensile stress and service life of pavement structure, the differential deformation rate of subgrade top under the frost heave of cohesive soil subgrade in the seasonal frozen area should be controlled within 2%.

5 Conclusions

Through a series of triaxial tests, finite element method and monitoring results, Mainly contents and conclusions are as following:

(1) Lateral displacement is growing with loading of soil filling, the more rapid the soil filling, the more rapid lateral deformation growing. To the toe of the new road, the biggest shift occurred in certain depth, the largest displacement was related to the soil thickness.
(2) Considering the overload load, the subgrade's differential deformation rate should be less than 3.07% and the differential deformation rate must be less than 2.27%.

(3) Considering the allowable bending tensile stress and service life of pavement structure, the differential deformation rate of subgrade top under the frost heave of subgrade in the seasonal frozen area should be controlled within 2%.

References

1. Cui, W., Lv, G., Liu, C.: Computational analysis of differential settlement on new subgrade in express way widening. Sci. Technol. Eng. 17(34) (2017)
2. Deng, W., Zhang, X., Chen, B.: Finite element analysis of influence of subgrade uneven settlement on asphalt pavement deformation. Acta Sin. Sin. 17(1) (2004)
3. Wang, H., Huang, X.: Finite element analysis of Expressway Widening on soft soil foundation. Highway Traffic Sci. Technol., 21(8) (2004)
4. Jiang, X., Qiu, Y.: Three dimensional finite element analysis of the whole process of old road widening. J. Eng. Geol. 13(03) (2005)
5. Zhang, D., Liu, S.: Parameter analysis of the influence of the construction of new embankment on the old embankment in the widening of expressway. Highway Traffic Sci. Technol. 23(7) (2006)
6. Su, C., Xu, Z.: Analysis method and engineering practice of foundation treatment design of expressway connection section. J. Eng. Geol. 8(1) (2000)
7. Huang, Q., Ling, J., Qian, J.: Influence of post construction differential deformation of new and old subgrade on pavement structure. J. Tongji Univ. (Nat. Sci. Ed.), 33(6) (2005)
8. Nie, P., Qu, X., Liu, F., et al.: Study on differential settlement index of subgrade after allowance for Shen Da expressway reconstruction and extension project. Highway Transport. Sci. Technol. 22(11) (2005)
9. Wang, B.: Study on Settlement Deformation Characteristics and Foundation Treatment Countermeasures of Expressway Splicing Section. Hehai University (2004)
10. Industrial Standard of the People's Republic of China. Technical Code for Highway Maintenance. Beijing People's Communications Press (1996)

Developing an Efficiency Equation for Tapered Pile Groups in Sand Using Mathematical and Numerical Analyses

Amin Shafaghat(✉), Hadi Khabbaz, and Behzad Fatahi

School of Civil and Environmental Engineering, University of Technology Sydney (UTS), PO Box 123 Broadway, Sydney, NSW 2007, Australia
Amin.Shafaghat@student.uts.edu.au, {Hadi.Khabbaz, Behzad.Fatahi}@uts.edu.au

Abstract. This study presents a new simple equation for prediction of pile group efficiency considering the effect of tapering angle in cohesionless soil under vertical loading. Firstly, a simple analytical relationship based on the mathematical definition of the pile group efficiency is developed. However, the effect of tapering angle is captured by defining a new geometry efficiency coefficient associated with the shaft vertical bearing component of tapered piles. Thereafter, a mathematical equation is developed by taking into account the shaft vertical bearing ratio and the new geometry efficiency coefficient. On the other hand, a numerical analysis is performed for modelling a single bored cylindrical pile and a tapered pile with the same volume as well as bored tapered pile groups to validate the proposed mathematical equation. The UBC sand constitutive model is adopted for the modelling of piles in loose Cambria sand. Subsequently, the load-displacement diagrams of single and group of piles are obtained. Then, the bearing capacities of cylindrical and tapered bored piles both as single and group are computed and compared, using a specific settlement criterion. Besides, the friction resistance ratio and the shaft vertical bearing ratio are separated, applying numerical methods. Having calculated the ratios of various components of bearing capacity, pile group efficiencies can be obtained from both numerical and mathematical models. The results show that the proposed equation can predict the pile group efficiency incorporating the tapering angle as well as other influencing parameters as a simple and novel relationship.

Keywords: Pile group · Efficiency coefficient · Load capacity · Settlement · Finite element model · Axial loading

1 Introduction

Capturing the behaviour of pile groups using numerical modelling, needs a three dimensional analysis which is a time-consuming effort. Hence, proposing an equation for predicting the bearing capacity of piles considering the important parameters such as the slenderness ratio, the pile spacing, the tapering angle, the sand internal friction angle

S. S. Kim et al. (Eds.): GeoChina 2021, SUCI, pp. 16–30, 2021.
https://doi.org/10.1007/978-3-030-80152-6_2

and the number of piles, is of great significance. Although, several group efficiency equations have been proposed to date, to the best of authors' knowledge, the effect of tapering angle has not been considered in the developed equations. On the other hand, most of the investigations on tapered piles are specified for isolated single piles rather than pile groups, relating to the complication assessment of pile group behaviour. The two main problems associated with pile group design are the pile group efficiency (η) and the settlement factor (S_f), which are defined as follows:

$$\eta = \frac{Q_g}{\sum Q_s} \tag{1}$$

$$S_f = \frac{S_g}{S_S} \tag{2}$$

where, Q_g is the capacity of the pile group, Q_s is the capacity of a single pile, S_g is the settlement of the pile group and S_s is the settlement of an individual pile.

According to Eq. (1), if the group capacity is equal to the sum of all individual pile capacities, then the group efficiency (η) will be equal to 1. The ASCE Committee on Deep Foundations (CDF) suggests that friction piles in cohesionless soils embedded at the usual spacing of s = 2D to 3D can have a group efficiency $\eta > 1$ (Committee, 1984). The reason assumed is that in frictional soil, the pile displacement along with driving vibrations increases the soil density in a vicinity zone of the pile. This effect is more significant in driven piles. However, considering friction piles in cohesive soils, the group capacity can be obtained as the summation of the point bearing along with block shear of the group. In this condition, the group capacity rarely can be more than the summation of single piles capacities.

The presence of pile cap also plays a significant role in pile group capacity as well as the pile group efficiency. For instance, if the pile cap is laying on the ground, the pile group will settle with the soil as the piles will also settle that much. For a free standing cap, the group bearing capacity of cylindrical piles would be the sum of the individual point capacities and block perimeter shear. While, for tapered piles, an additional shaft vertical bearing component should be taken into account. Hence, for tapered pile groups, the corresponding bearing ratio will affect the group efficiency, particularly for piles having larger angles (Shafaghat and Khabbaz 2020b).

Most of the developed relationships and guidelines are regarding to straight-sided wall piles. However, the attempt of this study is to investigate the characteristics of tapered piles featuring the effect of tapering angle. For this purpose, based on the mathematical definition of group efficiency factor, a new formula for a group of tapered piles is presented. The predicted relationship can be a function of a number of important parameters, as expressed in Eq. (3).

$$\eta = f(m, n, s, L, d, \alpha, \phi) \tag{3}$$

In the above equation, m is the number of piles in a row of a group, n is the number of piles in a column of a group, S is the spacing between piles, d is the average diameter of piles (equivalent to the same volume counterpart straight-sided wall pile diameter), α is the tapering angle of piles, L is the length of piles, and ϕ is the internal friction angle of the granular soil.

Afterwards, an array of numerical analyses is performed to model the single bored cylindrical and tapered piles as well as pile groups with the same volume of material. Thereafter, the load-displacement diagrams of single and group of tapered piles are obtained through numerical survey and the mathematical developed efficiency relationship is verified. The constitutive model for sand, developed by the University of British Columbia, known as UBC sand constitutive model (Beaty & Byrne 2011), is used for the numerical analysis of piles in loose Cambria sand.

2 Current Group Efficiency Equations

There are several relationships to predict the group efficiency for conventional cylindrical piles. Some relationships are established for group of piles in sand and some can predict the efficiency in clay. Besides, based on the pile cap condition of pile group, the efficiency coefficients can be divided into two categories including, pile group with a cap laying on the ground and pile group with a free-standing cap. Table 1 presents an overview at some of the well-known efficiency equations for cylindrical pile foundations. Most of the equations provide the efficiency of pile groups with a free-standing cap. As can be seen, the effect of tapering angle has not been taken into consideration in obtaining the pile group efficiency.

Table 1. Different methods for predicting the pile group efficiencies

Methods	Equations	Parameters	Applicable soil types
Terzaghi et al. (1996)	$Q_g = q_b BL + D_f(2B + 2L)S$	- q_b (ultimate capacity per unit area of a rectangular loaded area with dimensions B × L × D_f) - Q_g (the ultimate capacity of pile group) - B (width of pile group) - L (length of pile group) - S (average shearing resistance of soil per unit area)	Cohesive soils (soft clay)
Converse-Labarre formula, Bolin (1941)	$\eta_g = 1 - \frac{\xi}{90^\circ} \cdot \left[\frac{(n-1)\cdot m + (m-1)\cdot n}{m \cdot n}\right]$	- m (number of piles in a row) - n (number of rows) - $\xi = \arctan\left(\frac{d}{s}\right)$ in degrees - d (the pile diameter) - s (the centre-to-centre pile spacing)	Cohesive and granular soils

(continued)

Table 1. (*continued*)

Methods	Equations	Parameters	Applicable soil types
Los Angeles Group Action, Das (2004)	$\eta = 1 - \frac{D\left[m(n-1)+n(m-1)+(n-1)(m-1)\sqrt{2}\right]}{\pi Smn}$	- m (number of piles in a row) - n (number of rows) - D (the pile diameter) - S (the centre-to-centre pile spacing)	Cohesive soils
Feld (1943)	This method considers reducing the bearing capacity of single piles in the group by 1/16 so that the effect of each neighbouring pile in the same row is considered	- Pile arrangement	Cohesive and granular soils
Seiler and Keeney (1944)	$\eta = 1 - \frac{11S(n+m-2)}{7(S^2-1)(n+m-1)} + \frac{0.3}{n+m}$	- m (number of piles in a row) - n (number of rows) - S (the centre-to-centre pile spacing in ft)	Cohesive and granular soils
Poulos and Davis (1980)	$\frac{1}{\eta_g^2} = 1 + \left[\frac{(n \cdot m)^2 \cdot Q_O^2}{Q_B^2}\right]$	- Q_B (the ultimate load capacity of the block of pile) - Q_O (the ultimate load capacity of a single pile) - m (number of piles in a row) - n (number of rows)	Cohesive and granular soils
Pressure-area method, Chellis (1969)	$\eta = 1 - \frac{(2k+n)[2-(2k+n)]}{(2kn)^2}$	- n (the number of piles in each row), - $k = \frac{S'}{S}$ - S (The pile spacing) -S' (defined in Fig. 1)	Cohesive and granular soils
Kishida (1965)	This method is based on a number of diagrams	- Pile spacing - Soil friction angle - Pile diameter	Granular soils
Vesic (1967)	This method is based on a number of diagrams	- Pile spacing - Pile diameter	Granular soils
Sayed and Bakeer (1992)	$\eta = 1 - \left(1 - \eta' \cdot K\right) \cdot \rho$	-ρ (the friction factor), - K (the interaction factor) - η' (the geometric efficiency $\eta' = \frac{P_g}{\sum P_p}$) - P_g (the perimeter of the pile group) - $\sum P_p$ (sum of the perimeters of individual piles)	Cohesive and granular soils

(*continued*)

Table 1. (*continued*)

Methods	Equations	Parameters	Applicable soil types
Das (2004)	$\eta = \frac{2S(n+m-2)+4D}{nm}$	- m (number of piles in a row) - n (number of rows) - D (the pile diameter) - S (the centre-to-centre pile spacing)	Cohesive and granular soils
Hanna et al. (2004)	This method is based on the artificial neural network approach	- Pile arrangement - Pile spacing - Pile diameter - Pile installation method - Soil condition	Granular soils

Referring to Table 1, the most common parameters in predicting the pile group efficiency are the pile spacing, the pile diameter and the number of piles in each row and column. Even though, other factors, such as the interaction factor between the piles within a group, play a significant role in obtaining the efficiencies.

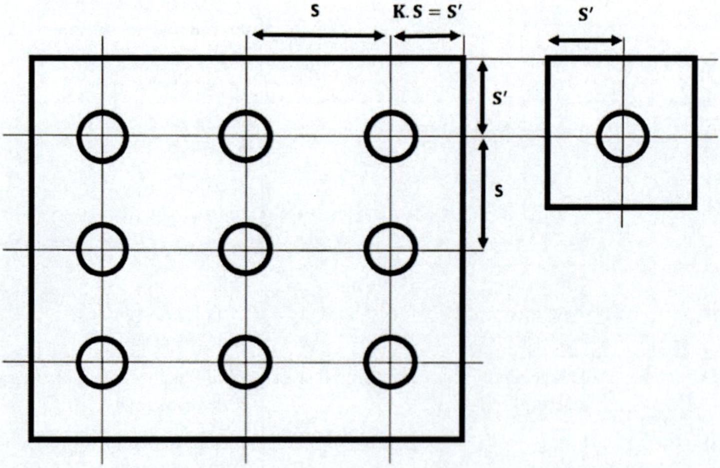

Fig. 1. Distribution of vertical stress under an individual pile and pile group (pressure-area formula) (after Chellis (1969)).

Figure 1 shows the vertical stress distributions under a 3 × 3 pile group and a single pile. It also demonstrates the definition of the parameter s' in the efficiency equation proposed by Chellis (1969) as presented in Table 1. Several other equations have been proposed to predict the efficiency of pile groups considering different parameters (Randolph 1994; Rao 2010; Tuan 2016; Zhao and Stolarski 1999). However, these relationships are only applicable for straight-sided wall pile groups and to the best of authors' knowledge, none of the above equations can capture the effect of all above-mentioned parameters simultaneously.

3 Analytical Approach

The analytical solution is routed from the basic definition of efficiency coefficient, which is as follows:

$$\eta = \frac{Q_g}{\sum Q_s} \tag{4}$$

where, Q_s is the bearing capacity of single pile and Q_g is the bearing capacity of the group. The load Q_s is conventionally divided into three components of shaft (skin) resistance (Q_f), the base (tip) resistance (Q_b), and the vertical component of bearing along the length of tapered pile due to inclination of their body (Q_{sv}). Figure 2 illustrates a tapered pile group along with separated resistance forces.

$$Q_s = Q_f + Q_b + Q_{sv} \tag{5}$$

Fig. 2.Tapered pile group and the separated resistance forces

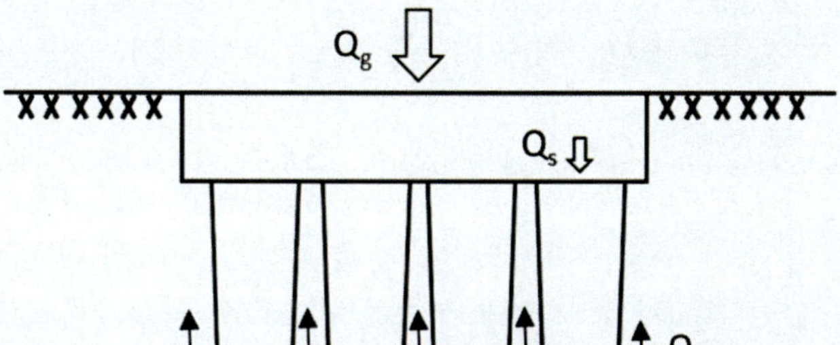

Fig. 2. Tapered pile group and the separated resistance forces

Vesic (1967) and Chellis (1969) have suggested that for piles embedded in sand, the group effect can be taken into consideration only for the shaft bearing component. According to this assumption, the group bearing capacity can be written as:

$$Q_g = \sum Q_b + \eta_{sv} \cdot \sum Q_{sv} + \eta_s \cdot \sum Q_f \tag{6}$$

where, the parameter η_s is the shaft load efficiency which is similar to what has been suggested by Sayed and Bakeer (1992), but with a slight modification to consider the effect of tapering angle. In addition, η_{sv} is the shaft vertical load efficiency, which is derived similar to the former load efficiency and can be written as:

$$\eta_s = \eta_s^{'} \cdot K \tag{7}$$

$$\eta_{sv} = \eta_{sv}^{'} \cdot K \tag{8}$$

where, $\eta_s^{'}$ is the geometric efficiency coefficient and K is the group interaction factor. Substituting Eqs. (6) to (8) in Eq. (4) yields:

$$\eta_g = \frac{\eta_s^{'} \cdot K \cdot \sum Q_f}{\sum Q_s} + \frac{\eta_{sv}^{'} \cdot K \cdot \sum Q_{sv}}{\sum Q_s} + \frac{\sum Q_b}{\sum Q_s} \tag{9}$$

Vesic (1980) has recommended that for a group having m × n piles, the ratio of $\frac{\sum Q_f}{\sum Q_s}$ is equal to the ratio of $\frac{Q_f}{Q_s}$ for a single pile. Hence, by assuming that the ratio of $\frac{\sum Q_{sv}}{\sum Q_s}$ is equal to the ratio of $\frac{Q_{sv}}{Q_s}$ for a single pile in sandy soil, the efficiency equation for a pile group in sand can be written as,

$$\eta_g = 1 + \left(\eta_s^{'} \cdot K - 1\right) \cdot \frac{Q_f}{Q_s} + \left(\eta_{sv}^{'} \cdot K - 1\right) \cdot \frac{Q_{sv}}{Q_s} \tag{10}$$

In addition, $\eta_s^{'}$ is defined in Eq. (11) (Sayed & Bakeer 1992) and $\eta_{sv}^{'}$ is derived similar to the friction geometric efficiency coefficient as presented in Eq. (12).

$$\eta_s^{'} = \frac{P_g}{\sum P_s} \tag{11}$$

$$\eta_{sv}^{'} = \frac{A_p}{\sum A_b + \sum A_{sv}} \tag{12}$$

where, P_g is the perimeter of the pile group, A_p is the area of the end surface of the block of pile group (as an equivalent large pile), $\sum P_s$, $\sum A_b$, and $\sum A_{sv}$ are the summations of the single piles lateral perimeter, the summation of base area of single piles, and the summation of shaft horizontal projected area corresponding to the vertical load bearing of single piles, respectively, as illustrated in Fig. 3. Accordingly, knowing that $A_{sv} + A_b = \frac{\pi}{4} D_{top}^2$ and for a circular pile group, $\eta_s^{'}$ and $\eta_{sv}^{'}$ can be defined as:

$$\eta_s^{'} = 2 \times \left[\frac{[(m-1) \times s + D_{av}] + [(n-1) \times s + D_{av}]}{\pi \times n \times m \times D_{av} \times \cos(\alpha)} \right] \tag{13}$$

$$\eta_{sv}^{'} = 4 \times \left[\frac{[(m-1) \times s + D_{bot}] \cdot [(n-1) \times s + D_{bot}]}{\pi \times n \times m \times D_{top}^2} \right] \tag{14}$$

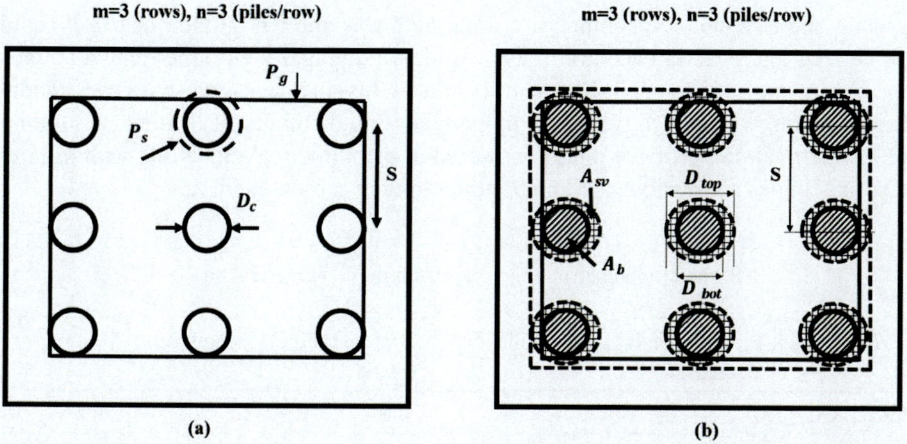

Fig. 3. Bottom view of a typical pile group pattern and the geometry efficiency coefficients definition a) shaft geometry efficiency (η_s') b) shaft vertical bearing geometry efficiency (η_{sv})

Hence, the final mathematical model for the efficiency of bored tapered pile groups in sand can be derived by Eq. (15).

$$\left\{h_g = 1 - \left(1 - 2 \times \left[\frac{[(m-1)\times s + D_{av}] + [(n-1)\times s + D_{av}]}{\pi \times n \times m \times D_{av} \times \cos(\alpha)}\right] \cdot K\right) \cdot \frac{Q_f}{Q_s} - (1 - 4 \times \left[\frac{[(m-1)\times s + D_{bot}] \cdot [(n-1)\times s + D_{bot}]}{p \times n \times m \times D_{top}^2}\right] \cdot K) \cdot \frac{Q_{sv}}{Q_s}\right\} \tag{15}$$

where, K can be obtained through existing data reported by Sayed and Bakeer (1992). Equation (15) gives the pile group efficiency by considering pile geometry parameters, including tapering angle and the sandy soil properties. It should be noted that for finding the total bearing capacity of piles using load-displacement diagrams, the load on pile heads corresponding to $0.05D_{av}$ of pile settlement (which is 75 mm) is used. Moreover, for a single pile (assuming $m = n = 1$), the geometry efficiency coefficients should be modified as the pile cross-sections were assumed to be circular, but the arrangements assumed to be square. While, for a single pile the geometry efficiency coefficients must be equal to 1. It is also worth mentioning that by increasing the spacing to large values, the group interaction factor approaches to zero. Hence, the group interaction factor has an inverse relationship with the pile spacing. Hence, the terms $\eta_s' K$ and $\eta_{sv}' K$ approach 1 for large spacing values.

4 Three-Dimensional Finite Element Modelling and Overview of the Numerical Models

The numerical modelling conducted in this study includes two sets of pile group models and two sets of single pile models with different geometries, but identical volume of material. Piles were modelled using a finite element software package, PLAXIS 3D (Brinkgreve et al. 2002), and the interface elements were used for shaft and toe of piles. In this analysis, piles with circular cross-section were embedded in an elasto-plastic ground and a monotonic compressive axial load was applied on their heads. It should be noticed that piles in each group were loaded individually with the same amount of loading as single pile models. Hence the results of this study can represent a free standing cap situation, where each pile in group partakes equal amount of loading. A summary of geometrical properties of piles considered in the numerical modelling with different tapering angles and various top and bottom radii are presented in Table 2.

Table 2. Geometry of the piles used in the numerical analyses

L/D	Piles	Tapering angle (degrees)	Top radius (*mm*)	Bottom radius (*mm*)	Length (*m*)	Volume (m^3)
10	CL	0	750	750	15	26.5
	TL	1.4	925	559	15	26.5

The enhanced meshing system was applied surrounding each pile, where the pile and soil interact through the interface plane as shown in Fig. 4. This refinement could improve the mesh quality and consequently contribute to more accurate results. Piles were modelled as solid elements and volumetric objects to consider the tapering effect and to capture the stress states surrounding them.

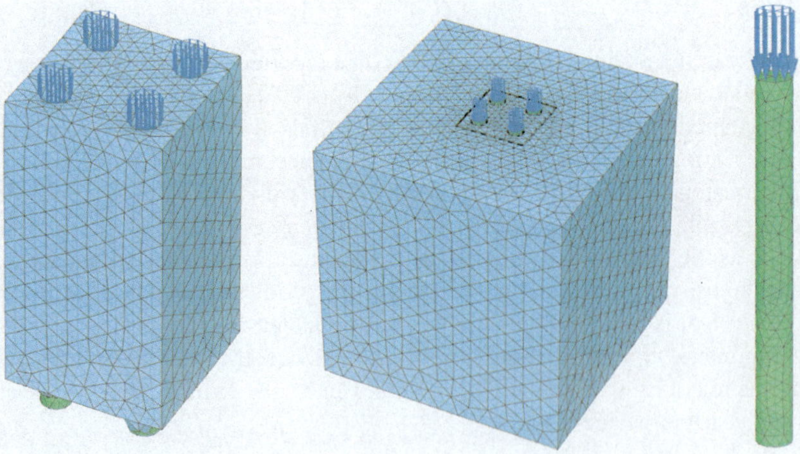

Fig. 4. Enhanced meshed system used for piles in 3D numerical analyses

The interface elements between pile and the adjacent soil were considered and the reduction factor (R_f) of 0.7 was used. This factor applies to the strength and stiffness parameters of soil as Eqs. (16) and (17) show.

$$tan\phi_i = R_{inter}\ tan\phi_{soil} \tag{16}$$

$$G_i = R_{inter}^2 \cdot G_{soil} \tag{17}$$

In the above equations, ϕ_i is the internal friction angle of the interface, ϕ_{soil} is the internal friction angle of the soil, G_i is the shear modulus of the interface, and G_{soil} is the shear modulus of the soil. Figure 5 illustrates the three-dimensional model of piles embedded in soil along with the interfaces surrounding the piles as well as interfaces around pile group blocks.

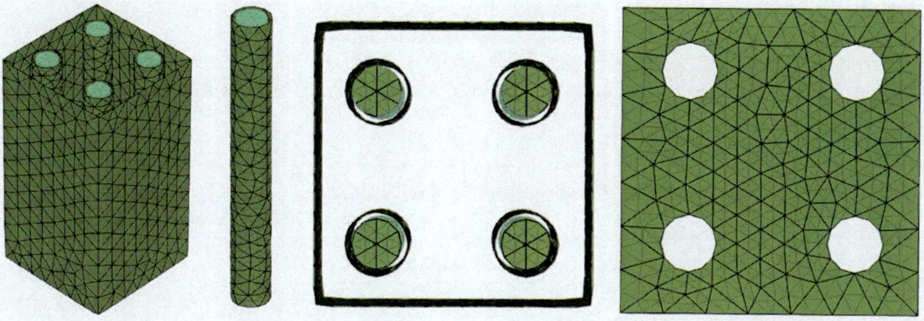

Fig. 5. Interfaces surrounding the individual piles and at the lateral and bottom faces of pile blocks

5 Validation of the Obtained Efficiency Equation Using 3D FEM Numerical Analysis

The load-displacement behaviour of modelled piles are considered as reference data to calculate the bearing capacities (Q_f, Q_{sv}, Q_s) using numerical integration of vertical effective stress on the base surface area of piles (Hataf & Shafaghat 2015a, 2015b; Shafaghat & Khabbaz 2020a; Shafaghat et al. 2018). Therefore, these load-displacement diagrams for a single pile and group of piles are extracted as a result of three dimensional numerical modelling, as shown in Figs. 6 and 7. Both single piles and pile groups with an arrangement of 2 × 2 and slenderness ratio of $\frac{L}{D_c} = 10$ (where D_c is the diameter of the reference same volume cylindrical pile) are modelled in loose Cambria sand. It can be noted that the UBC sand model is used for the numerical analysis. The pile model material parameters including the Poisson's ratio, Young's modulus, unit weight and its behavioural model are presented in Table 3, respectively. The spacing of $S = 3D_{top}$ is used to model the piles in groups for this study as it is perceived to be the optimum spacing of piles to have efficient interaction in group (Poulos 1979; Poulos & Davis 1980). The reference settlement of piles for interpreting different components of bearing capacity is assumed to be as $\delta_{ref} = 0.05D_{av}$. Moreover, concerning the group interaction factor (K), which is a function of pile spacing, the method of pile installation, and type of soil, its values ranging from 0 to 1 (Hanna et al. 2004; Sayed & Bakeer 1992). According to the relative density of the target sand, and based on the recommendations in the literature (Hanna et al. 2004), the group interaction factor (K) for a 2 × 2 bored pile group, embedded in loose sand, is $K = 0.5$.

Table 3. Pile material parameters

Parameters	$\gamma\left(\frac{kN}{m^3}\right)$	ν	E (GPa)	Material model
Values	24	0.15	25	Linear-elastic

Table 4 represents the input parameters of soil properties used in the numerical modelling based on the UBC sand model.

Table 4. Sandy soil properties adopted for numerical analysis based on UBC sand model (loose Cambria sand)

k_B^e	k_G^e	k_G^p	me	ne	np	ϕ_{cv}	ϕ_p	c	R_f	f_{dens}	$\gamma_{unsat}\left(\frac{kN}{m^3}\right)$	$\gamma_{sat}\left(\frac{kN}{m^3}\right)$	$(N_1)_{60}$
150	300	330	0.25	0.25	0.25	31°	32°	0	0.98	0.3	15.3	19.3	5

In Table 4, k_B^e is the elastic bulk modulus factor, k_G^e is the elastic shear modulus factor, k_G^p is the plastic shear modulus factor, *me* is the rate of stress-dependency of elastic bulk modulus, *ne* is the rate of stress-dependency of elastic shear modulus, np is the rate of stress-dependency of plastic shear modulus, ϕ_{cv} is the constant volume friction angle, ϕ_p is the peak friction angle, c is the cohesion, R_f is the failure ratio, f_{dens} is the densification factor and $(N_1)_{60}$ is the corrected SPT value.

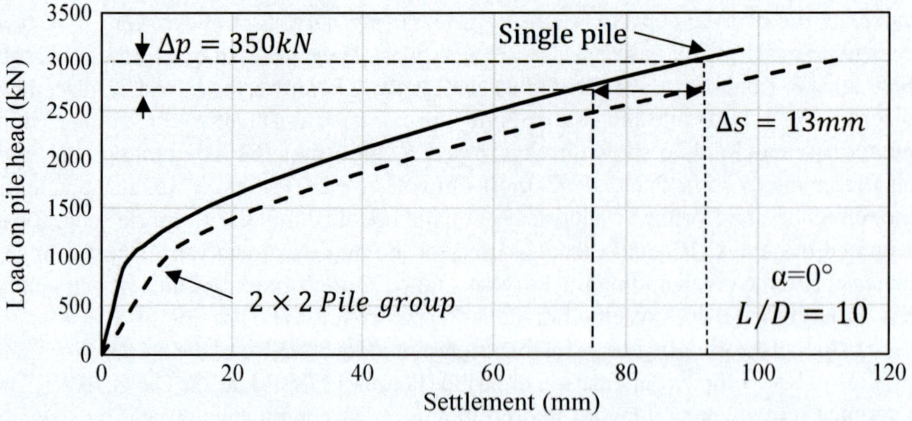

Fig. 6. Load-displacement curves for single and group of cylindrical pile modellings (Loose Cambria sand)

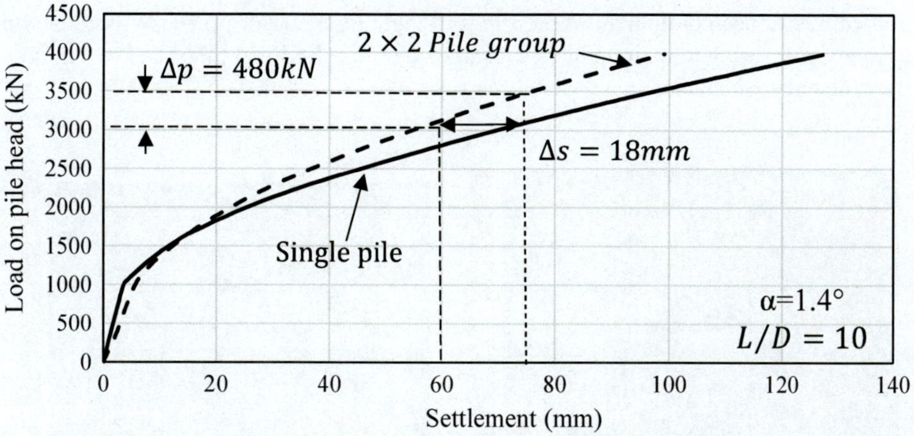

Fig. 7. Load-displacement curves for single and group of tapered pile modellings (Loose Cambria sand)

According to the load-displacement results of Figs. 6 and 7 obtained from the three-dimensional numerical modelling, and based on the numerical calculation techniques used to find the various components of resistances (Hataf & Shafaghat 2015a), the friction bearing ratio and the shaft vertical bearing ratio are obtained and presented in Table 5.

Table 5. Bearing capacity ratios of a single same volume cylindrical and tapered piles

	Cylindrical pile	Tapered pile with α = 1.4°
Shaft bearing ratio ($\frac{Q_f}{Q_s}$)	0.23	0.27
Shaft vertical bearing ratio ($\frac{Q_{sv}}{Q_s}$)	0	0.35

According to Figs. 6 and 7 and using Eq. (1) the pile group efficiencies of both 2 × 2 cylindrical and tapered piles are 0.90 and 1.1, respectively. On the other hand, using Eq. (15), the pile group efficiencies are obtained 0.91 and 1.06, which shows the presented efficiency equation can simply predict the pile group efficiency while considering the tapering angle with a reasonable accuracy. Figures 8 and 9 illustrate the relative shear stress state and the vertical displacement of piles at the end of the block of cylindrical and tapered groups, respectively.

Figure 9 illustrates that the vertical displacement of the soil at the centre part of the tapered pile group is more uniform comparing to the vertical displacement of the cylindrical pile group. This is due to the shaft inclination of tapered piles which makes the upper cross-section bigger than the toe. In other words, the tapering effect can compact the surrounding soil downward, which in turn densifies the adjacent soil while the pile experiences incremental settlement. Hence, for tapered piles embedded in sand, the lateral earth pressure coefficient will increase by each incremental settlement of pile. This will lead to increase in both the shaft friction resistance as well as the shaft vertical resistance component. Accordingly, single tapered piles provide more bearing capacity comparing to their counterpart same volume cylindrical piles, and will improve the efficiency when used in a group.

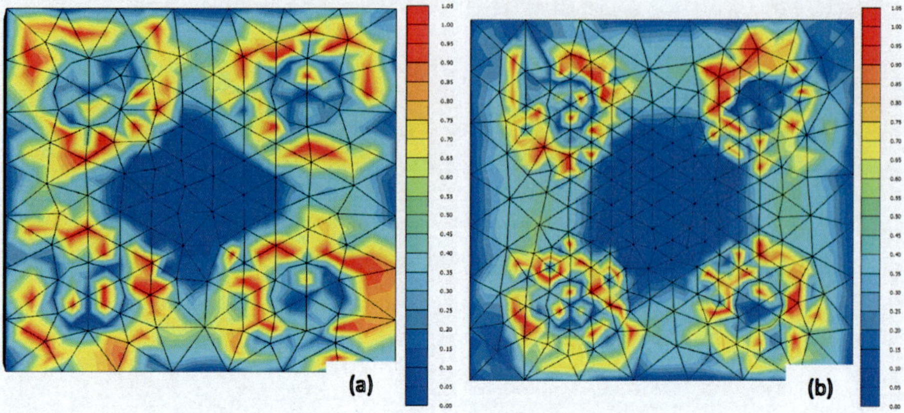

Fig. 8. Relative shear stress state on the end surface of the block of pile groups (bottom view), a) cylindrical pile group, b) tapered pile group ($\alpha = 1.4°$)

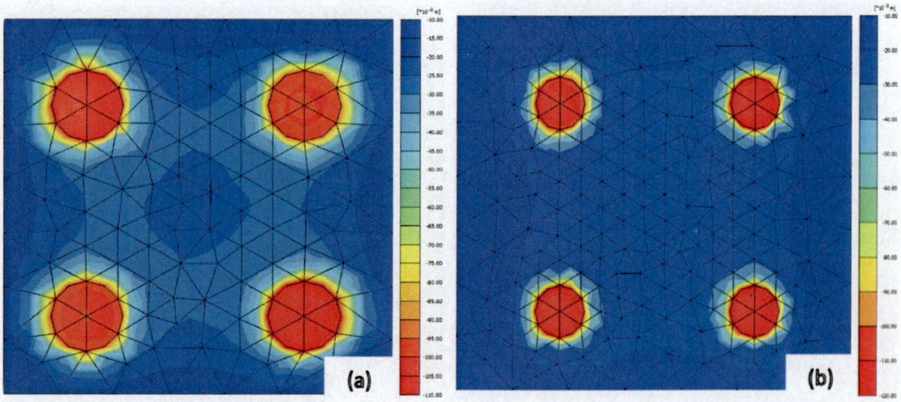

Fig. 9. Vertical displacement state on the end surface of the block of pile groups (bottom view), a) cylindrical pile group, b) tapered pile group ($\alpha = 1.4°$)

6 Conclusions

A simple equation to obtain the group efficiency of bored piles incorporating the tapering angle has been presented. According to the efficiency values for a 2 × 2 pile group, it is obvious that the tapering angle can significantly affect the bearing capacity of piles when arranged in a group. If the tapering angle increases from 0° to 1.4°, the efficiency will enhance more than 15%.

The load-displacement diagrams of single piles in the elastic zone represent a more stiff behaviour than piles in a group (for settlements less than $0.01D_{av}$). However, for a group of tapered piles and for settlements more than $0.01D_{av}$, the load-displacement diagram locates above the load-settlement diagram of a single pile, which in turn can contribute to higher efficiency values.

As can be seen, by increasing the tapering angle, the friction bearing and the shaft vertical bearing ratios increase. This increase is mainly due to the decrease in the end surface area of the piles and the increase of shaft horizontal projected area. Accordingly, there should be an increase in the group interaction factor by increasing the tapering angle. Field test data is required to investigate this finding more precisely.

A new geometry efficiency coefficient has been proposed for the shaft vertical bearing component of piles. This new coefficient plays a significant role in predicting the pile group efficiency as this bearing component increases by increasing the tapering angle. Hence, the efficiency equation can be decomposed into two separate components including the friction efficiency and the vertical bearing efficiency.

Similarly, an equation can be developed for the settlement factor of pile groups with considering the effect of tapering angle and slenderness ratio. The proposed equations can be employed by practicing engineers to predict the behaviour of cylindrical and tapered pile groups embedded in sand in a simple and time saving manner.

References

Beaty, M.H., Byrne, P.M.: UBCSAND Constitutive Model Version 904aR. University of British Columbia (2011)

Bolin, H.: The pile efficiency formula of the Uniform Building Code. Build. Stand. Month. **10**(1), 4–5 (1941)

Brinkgreve, R., Swolfs, W., Engine, E.: Plaxis users manual. Balkema, Rotterdam, The Netherlands (2002)

Chellis, R.D.: Pile Foundations, 2nd ed McGraw-Hill Inc., New York, NY (1969)

ADF Committee: Practical Guidelines for the Selection, Design and Installation of Piles, vol. 5, p. 7. ASCE Deep Foundation Committee, USA (1984)

Das, B.: Principles of Foundation Engineering, p. 489. Brooks/Cole-Thomson Learning Inc., USA (2004)

Feld, J.: Discussion on friction pile foundations. Trans. Am. Soc. Civ. Eng. **108**, 143–144 (1943)

Hanna, A.M., Morcous, G., Helmy, M.: Efficiency of pile groups installed in cohesionless soil using artificial neural networks. Can. Geotech. J. **41**(6), 1241–1249 (2004)

Hataf, N., Shafaghat, A.: Numerical comparison of bearing capacity of tapered pile groups using 3D FEM. Geomech. Eng. **9**(5), 547–567 (2015)

Hataf, N., Shafaghat, A.: Optimizing the bearing capacity of tapered piles in realistic scale using 3D finite element method. Geotech. Geol. Eng. **33**(6), 1465–1473 (2015)

Kishida, H.: Bearing capacity of pile groups under eccentric loads in sand. In: Proc. 6th ICSMFE, Montreal, vol. 2, pp. 270–274 (1965)

Poulos, H.G.: Group factors for pile-deflection estimation. J. Geotech. Geoenviron. Eng.*105* (ASCE 15032) (1979)

Poulos, H.G., Davis, E.H.: Pile Foundation Analysis and Design (1980)

Randolph, M.: Design methods for pile group and piled rafts. In: Paper presented at the Proc. 13th Int. Conf. on SMFE (1994)

Rao, N.K.: Foundation Design: Theory and Practice. John Wiley & Sons (2010)
Sayed, S.M., Bakeer, R.M.: Efficiency formula for pile groups. J. Geotech. Eng. **118**(2), 278–299 (1992)
Seiler, J., Keeney, W.: The efficiency of piles in groups. Wood Preserv. News **22**(11), 109–118 (1944)
Shafaghat, A., Khabbaz, H.: Numerical evaluation of bearing capacity of step-tapered piles using P-Y curves analysis. In: Shehata, H., Badr, M. (eds.) Advancements in Geotechnical Engineering: The official 2020 publications of the Soil-Structure Interaction Group in Egypt (SSIGE), pp. 200–212. Springer International Publishing, Cham (2021). https://doi.org/10.1007/978-3-030-62908-3_15
Shafaghat, A., Khabbaz, H.: Recent advances and past discoveries on tapered pile foundations: a review. Geomech. Geoeng. 1–30 (2020b)
Shafaghat, A., Khabbaz, H., Moravej, S., Shafaghat, A.: Effect of footing shape on bearing capacity and settlement of closely spaced footings on sandy soil. Int. J. Geotech. Geol. Eng. **12**(11), 676–680 (2018)
Terzaghi, K., Peck, R.B., Mesri, G.: Soil Mechanics in Engineering Practice. John Wiley & Sons (1996)
Tuan, P.: A simplified formular for analysis group efficiency of piles in granular soil. Int. J. Sci. Eng. Res. **7**(7), 15–21 (2016)
Vesic, A.: Predicted behavior of piles and pile group at the Houston site. In: Paper presented at the Proc. Pile group Prediction Symp (1980)
Vesic, A.S.: A Study of Bearing Capacity of Deep Foundations. Engineering Experiment Station (1967)
Zhao, Y.J., Stolarski, H.K.: Stability of Pile Groups. Technical report documentation, No. MN/RC-31. University of Minnesota (1999)

Analysis of Swelling Shrinkage Cracks Development Effects in Improved Expansive Soil Using Image Processing Technology

Cheng Chen[1,2](✉), Jian-fei Liu[1], and Jun Gong[2]

[1] Hunan Engineering Technology Research Center for High-Speed Railway Operation Safety Assurance, Hunan Vocational College of Railway Technology, Hunan 412006, China
[2] College of Civil Engineering and Mechanics, Xiangtan University, Hunan 411105, China

Abstract. To study the development characteristic of the improved expansive soil fissure and its effect on the shear strength, the image processing technology is used to make a quantitative analysis on the surface of the expansive soil crack parameters after the dry-wet circulation with different modified methods. The results show that the soil surface crack rate increases with the number of dry-wet circulation, and tends to a stable state. Adding mixed sand or lime, it is found that the crack rate increases with the degree of compaction, and the effection is not the same. According to the rules of the development of the improved expansive soil surface crack, the main conclusion of this study is that combination of the fracture index is more reasonable to evaluate the improvement effect of the expansive soil.

Keywords: Improved expansive soil · Crack rate · Swelling and shrinkage crack · Fractal dimension

1 Introduction

Expansive soil is a high-plasticity clay mainly composed of clay minerals, and sensitive to changes in the environment's humidity and heat. The expansive soil slopes and embankment is subject to expansion and contraction deformation due to the dry-wet cycle, resulting in the development of fissures and the collapse of the soil slope [1–5]. The above situation occurs because of the initiation, extension and penetration of fissures, which destroys the integrity of the soil; on the other hand, because of the existence of fissures, it provides a good channel for rainwater infiltration and evaporation, resulting in the development of fissures of the soil. Due to the unevenness and variability of the fissures, the expansive soils exhibit different strength behavior [2]. To reveal the strength and deformation failure mechanism of expansive soil, it is found that the shear strength behavior, unsaturated soil mechanical theory, deformation behavior and crack development rules of expansive soil under the action of dry-wet cycle, as well as the engineering characteristics and treatment of expansive soil is of great significance.

With the application of fractal geometry theory, researchers have made many gratifying results from the perspective of the combination of geometric deformation and

S. S. Kim et al. (Eds.): GeoChina 2021, SUCI, pp. 31–39, 2021.
https://doi.org/10.1007/978-3-030-80152-6_3

the crack structure of expansive soils. The development of X-ray diffraction (XRD), scanning electron microscopy (SEM), transmission electron microscopy (TEM), and the application of digital image processing technology have further promoted the study of the micro-structure of expansive soil by researchers and engineers. Lu Zaihua et al. [3] obtained CT numbers and variance numbers of expansive soil samples using CT technology, discussed the evolution law of internal cracks of expansive soil samples during the dry-wet cycle, and quantitatively analyzed the influence of these two parameters. Yuan Junping [4] used an optical microscope to observe the dynamic change law of the cracks in the remolded expansive soil, and proposed that the gray level entropy can be used to better describe the development law of the cracks. Zhang Jiajun et al. [5] used vector graphics technology to vectorize the crack development map of expansive soil obtained after scanning, and described the evolution of the crack through the collected geometric parameters such as the average length of the crack, the average distance between the cracks and the total number of cracks. mechanism. Tang Chaosheng et al. [6] used digital image processing technology to observe the shrinkage and cracking of the expansive soil slurry during the drying process, and proposed that the expansive soil is prone to cracking when it loses water. Yang Heping [7], Huang Zhen, etc. [8] studied the influence of compaction degree on the development of expansive soil fissures through laboratory experiments, and used the particle and fissure image recognition and analysis system (PCAS) developed by Nanjing University to study the dry and wet cycles. Afterwards, the development law of the cracks on the surface of the expansive soil under different degrees of compaction was analyzed quantitatively.

At present, with the deepening research on the development of fissures in expansive soils, researchers have gradually focused on the improvement of expansive soils. And there are mainly three types of physical improvement, chemical improvement and biological improvement. Physical improvement includes techniques such as reinforcement, compaction, sand mixing, and water barrier; chemical improvement includes methods such as lime, lime-lime, and cationic additives. In this paper, combined with engineering practice, soil samples were taken on site, and laboratory experiments were conducted to analyze the influence of different compaction degrees on the development of expansive soil cracks. On this basis, combined with Matlab software image processing technology, obtained the variation law of the crack rate, quantitatively described the development degree of the crack, and studied the compaction degree, the effect mechanism of the amount of weathered sand and lime on the cracking of the expansive soil.

2 Laboratory Test

The sample soils of this experiment comes from the filling roadbed section of a highway in Xiangtan, Hunan. Its basic physical properties are shown in Table 1.

Table 1. Basic physical index

ρ(g/cm^3)	W_L%	I_P	w_{opt}%	$\rho_{d\,max}$(g/cm^3)	Particle composition			δ_{ef}%
					>0.075	0.075–0.005	< 0.005	
2.08	51	27	16.0	2.02	0.28	52.14	47.58	65

2.1 Sample and Testing Design

Taking the natural air-dried soil sample by quartile method, smash it with a hammer and pass through a 2 mm sieve . By adding a certain amount of pure water to the soil sample, configure the soil sample, keep its initial moisture content at around 18.0% of the natural moisture content, sealed and covered the damp material with a damp cloth for more than 24 h, to ensure that the moisture distribution in the soil sample is even. Taking a compaction cylinder, configured the reshaped soil sample into a sample with a dry density of 1.8 g/cm^3, and compacted the soil sample by static compaction. The initial saturation of the obtained sample was 65%.

In this experiment, six dry-wet cycles were set up. First, the soil sample was dehumidified from the initial moisture content to a constant weight of the sample, then purified water was added to increase the moisture content to near the initial moisture content, and then dehumidified again to a constant weight, that is, complete a dry and wet cycle.

2.2 Testing Procedure

(1) The test process is divided into two stages: First, considering three different degrees of compaction, and analysis the expansion and contraction characteristics and crack development rules under different degrees of compaction. Second, considering the effect of two different improved methods on the crack characteristics of expansive soils.
(2) In the first stage, keep the optimal moisture content of the sample unchanged, use the static pressure method to carry out the indoor compaction test, control the degree of compaction to be 80%, 85% and 90% respectively, and select the inner diameter of the test container, which is a rectangular parallelepiped glass container of 20 cm × 20 cm × 5 cm, and the thickness of the sample is 20 mm. A constant temperature oven at 70 °C is used to control the dehumidification process of the test soil sample. On the other hand, at a position about 1 m away from the sample, use a sprayer to sprinkle the test soil sample to increase the humidity. Finish to prevent the adverse effect of water spray on the surface of the sample. After the sample reaches saturation, it can be placed in a constant temperature and humidity box for more than 24 h. During the humidification process, ensure that the moisture content inside the soil is evenly distributed. In the process of dehumidification and humidification, a high-definition Canon digital camera is used, equipped with a special tripod, to take pictures of the test soil samples at a fixed height, and the Matlab on the PC side is used to set the camera's photographing interval to 20 min. At the same time, choose an opaque laboratory and place table lamps around the tripod to eliminate the effects of sunlight on the photo effects. When the amount of dehydration per hour is less than 0.2 g for 3 consecutive hours, it can be determined that the sample has tended to dry and constant weight, and the dehumidification process has been completed.
(3) The second stage is the improved expansive soil dry-wet cycle test, to maintain the test soil sample to the best degree of compaction, and add different amounts of weathered sand and lime to the sample. Six dehumidification and humidification tests under the same conditions were also conducted six times, and photographs

were taken to observe the development law of the cracks on the surface of the expansive soil improved by two different methods. The test container is the same as the first stage, the amount of weathered sand is selected as 5%, 15% and 20%, and the amount of lime is selected as 3%, 6% and 9%.

3 Image Binarization Processing

3.1 Binary Processing and Crack Vector

In order to observe the development law of expansive soil fissures more vividly, this paper uses the image processing function of Matlab software to write a program to carry out the image binarization processing on the expansion and contraction fissure pictures of the expansive soil samples. The principle is to convert the picture into black and white images with gray values of 0 and 1. In the binary image, black pixels represent cracks and are represented by "0"; white pixels represent expansive soil blocks and are represented by "1", the processing result is shown in Fig. 1.

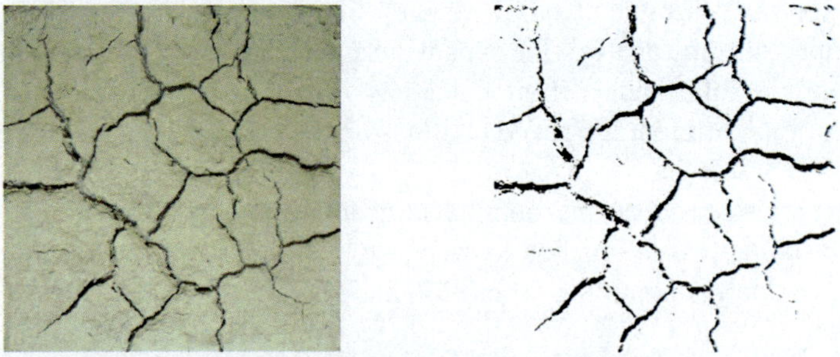

Fig. 1. Fissure pictures and vectors

3.2 Calculation of Crack Rate and Fractal Dimension

The area of the crack is related to the shrinkage of the soil body. The area of the crack in the surface area of the soil body per unit area is called the crack rate. It is expressed as the percentage value of the black pixel and the sum of the black pixel and the white pixel in the binary image. As shown in Eq. (1),

$$P = \frac{n_b}{n_w + n_b} \times 100\% \tag{1}$$

In which, n_b is the number of black pixels in the fissure, n_w is the number of white pixels in the cracked soil block of expansive soil.

In order to further analyze the development of the fissure structure morphology on the sample surface, another fissure index is introduced, namely the fractal dimension of the fissure. The fractal dimension can characterize the irregularity of complex shapes and reflect the effectiveness of the space occupied by complex shapes. Perform noise reduction processing on the black crack image in the image, remove the disturbing pixels, and then solve the fractal dimension of the image. The expression of the branch dimension D of the crack rate of the expansive soil is shown in Eq. (2),

$$D = \lim_{\varepsilon \to 0} \frac{\ln(N(\varepsilon)/a)}{\ln(1/\varepsilon)} \tag{2}$$

In which, a is a constant and ε is an arbitrary length value. In the crack zone of the sample, a square grid with side length is used to randomly cover, calculate the number of grids occupied by the crack during each coverage, and it can be calculated the size of the fractal dimension according to the Eq. (2).

4 Analysis of Test Results

4.1 Crack Development Rules Under Different Compaction Degrees

The test lasted 720 h. The sample passed through two stages, namely the surface shrinkage stage and the crack evolution stage. The shrinkage of the fracture surface shows that the specimen shrinks toward the center as a whole, while the evolution stage of the fracture shows the evolution of the fracture from the occurrence to the interpenetration. Under three different degrees of compaction, the evolution of the expansion soil cracks are shown in Fig. 2, Fig. 3 and Fig. 4, respectively.

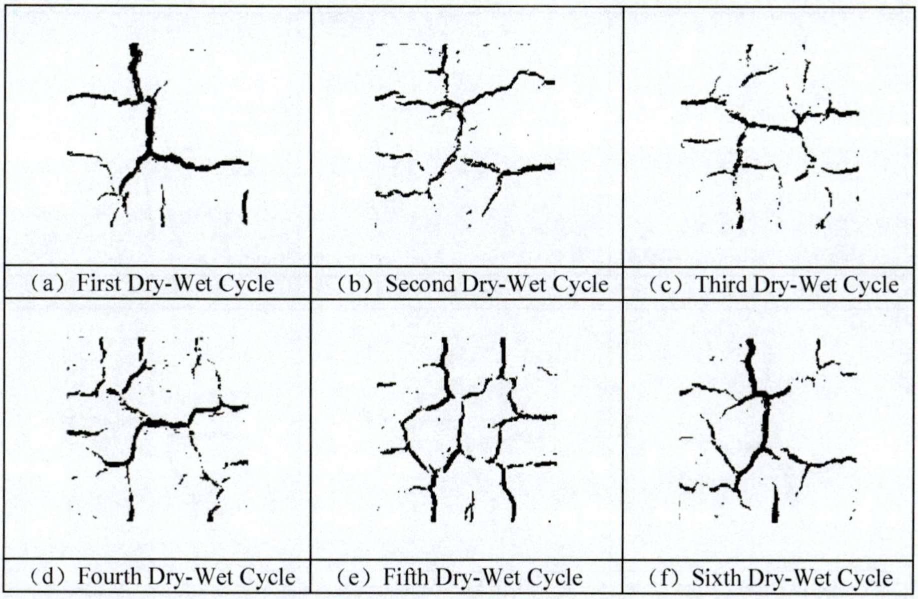

Fig. 2. Image of swelling and shrinking cracks of expansive soil under dry-wet cycle (80% compaction degree)

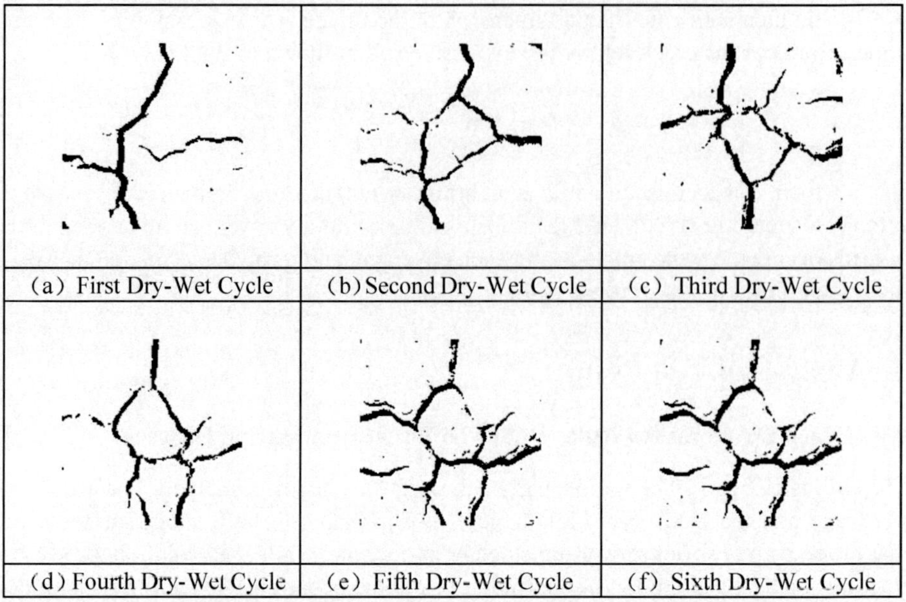

Fig. 3. Image of swelling and shrinking cracks of expansive soil under dry-wet cycle (85% compaction degree)

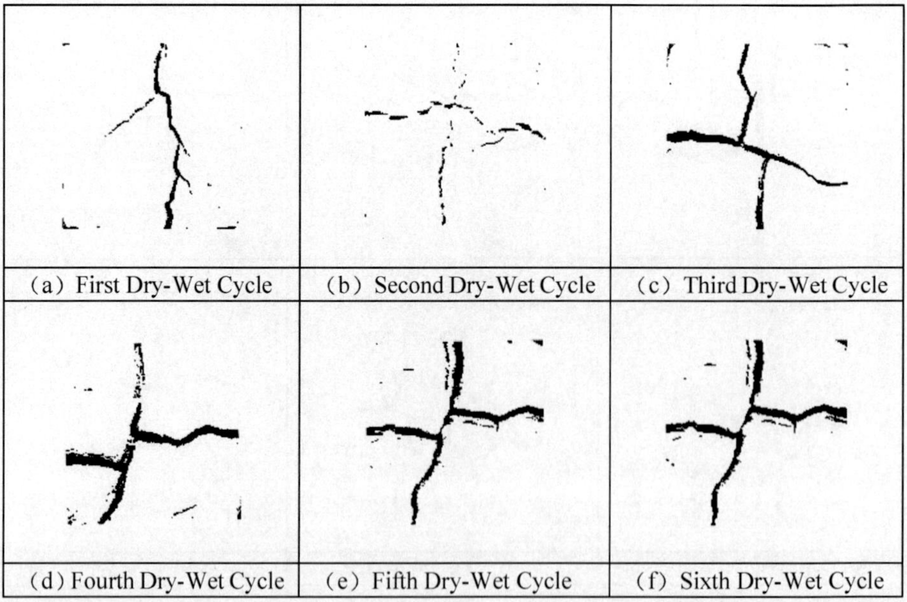

Fig. 4. Image of swelling and shrinking cracks of expansive soil under dry-wet cycle (90% compaction degree)

It can be seen from the fracture development of expansive soils with different degrees of compaction in Fig. 2, Fig. 3 and Fig. 4, that the degree of compaction has a greater influence on the development state of the fracture of expansive soil. When the sample only goes through a dry-wet cycle, with the increase of the degree of compaction, the degree of crack development of the expansive soil is weakened, the crack performance is finer, and the cracks are rarer. However, as the number of dry and wet cycles increases, the degree of cracking will continue to increase, the width of the crack will increase, and new cracks will continue to be generated, and the original cracks will gradually expand. By comparing the fifth and sixth dry and wet cycles, it was found that after a certain number of cycles, the development of fissures gradually stopped, new fissures no longer grew, and the width of fissures no longer increased. The test results show that increasing the degree of compaction can effectively reduce the growth of cracks, and the development of cracks on the surface of the sample has a certain continuation feature. Most of the crack developments extend along the original cracks, mainly widening and deepening. Both the fracture rate and fractal dimension increase with the increase of the number of dry and wet cycles, and eventually tend to a stable state.

4.2 Fracture Development Rules Mixed With Weathered Sand

The expansive soil is mixed with 5%, 10%, and 15% of weathered sand, and the compaction is kept at 90%. After six dry-wet cycles, the first and sixth dry-wet conditions are obtained. The crack development rules of the circulating sample is shown in Fig. 5.

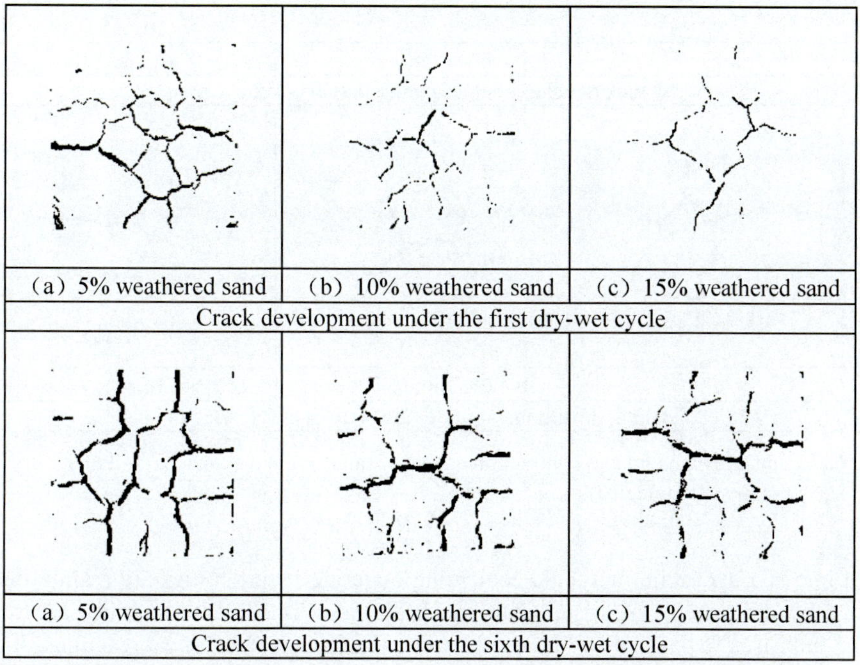

Fig. 5. Image of expansion and contraction cracks of sand-swelled expansive soil under dry-wet cycle (90% compaction degree)

Comparing Fig. 4 and Fig. 5, it can be seen that, compared with the development test results of expansive soil fissures without the addition of weathered sand, the addition of weathered sand can better suppress the development of expansive soil fissures, mainly because the main component of the weathered sand is two Silicon oxide and aluminum oxide, and the active ingredients of expansive soil are montmorillonite and illite, which are rich in calcareous elements. After the two are combined, the elements are prone to ion exchange between each other, generating gelling substances, such as calcium silicate and calcium aluminate, thereby producing a cementing effect, reducing the activity of expansive soil, effectively reducing its expansion and contraction force.

4.3 Crack Development Rules Mixed With Lime

Mixing 3%, 6% and 9% lime in the expansive soil respectively, keeping the compaction degree at 90% unchanged, after six dry-wet cycles, the first and sixth dry-wet cycles are obtained. The regularity of the crack development of the sample is shown in Fig. 6.

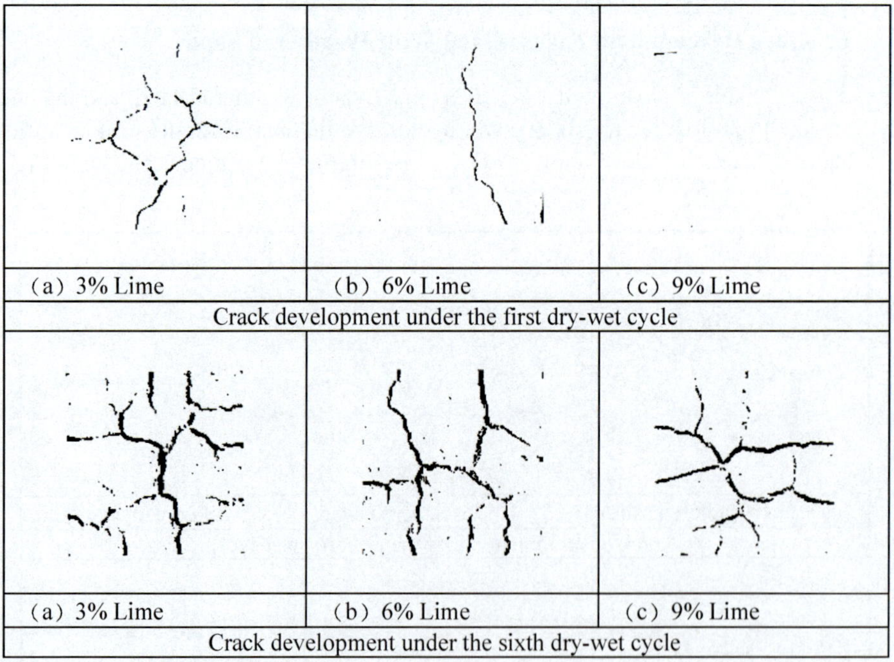

Fig. 6. Image of expansion and contraction cracks of sand-swelled expansive soil under dry-wet cycle (90% compaction degree)

Lime is a hydraulic inorganic cementing material, its properties are mainly determined by the content of active CaO and MgO, and its cementing ability is strong. Lime-improved expansive soil has both physical effects and chemical reactions. It can be seen from Figs. 4, 5 and 6 that after lime is added, the fissure development degree of the expansive soil is weaker than that without lime, and its improvement effect is even better than that of The weathered sand is better. At the same time, with the increase of the amount of lime, the cracks on the surface of the expansive soil develop weaker and weaker, and finally disappear almost, showing that the crack rate becomes smaller and smaller.

5 Conclusions

(1) During the first dehumidification process, the expansive soil sample only produced volume shrinkage and no cracks. With the increase of the number of dry-wet cycles, the cracks gradually developed, showing the increase in crack rate and fractal dimension, but with the release of the tensile stress potential energy of the soil, the crack development will gradually stabilize and eventually stop.
(2) The average width of the fissures and the total length of the fissures gradually increase with the number of dry-wet cycles, and eventually tend to a fixed value. The number of cracks also increases with the number of dry-wet cycles.
(3) The cracks of the improved expansive soil increase with the number of dry-wet cycles, and eventually tend to a stable state. The development of fractures mostly extends along the original fractures, which is mainly manifested by the widening and deepening of the original fractures.
(4) Increasing the degree of compaction, mixing with weathered sand and lime can effectively suppress the development of cracks on the surface of the expansive soil. Using reasonable weathering sand or lime to improve the compacted soil, which can enhance the stability of the project and avoid major safety hazards.

Acknowledgments. This study was financially supported by Hunan Natural Science Foundation of China (No: 2019JJ60010 and No: 2019JJ60070).

References

1. Yuan, J., Yin, Z., Bao, C.: Quantitative description method and index for fissure in expansive soils. J. Yangtze River Sci. Res. Inst. **20**(6), 27–30 (2003)
2. Li, X., Feng, X., Zhang, Y.: Depicting and analysis of expansive soil fissure in view of plane. Hydrogeol. Eng. Geol. **36**(1), 96–99 (2009)
3. Lu, Z.-H., Chen, Z.-H., Pu, Y.: A CT study on the crack evolution of expansive soil during and wetting cycles. Rock Soil Mech. **23**(4), 417–422 (2002)
4. Yuan, J., Yin, Z.: Quantitative index of fissure and strength characteristics of fissured expansive soils. J. Hydraul. Eng. **6**, 108–112 (2004)
5. Zhang, J., Gong, B., Hu, B., et al.: Study of evolution law of fissures of expansive clay under wetting and drying cycles. Rock Soil Mech. **32**(9), 2729–2734 (2011)
6. Tang, C., Wang, D., Shi, B.: Quantitative analysis of soil desiccation crack network. Chin. J. Geotech. Eng. **35**(12), 2298–2305 (2013)
7. Yang, H., Lliu, Y., Li, H.: The development of cracks of compacted expansive soil under dry-wet cycling. J. Transport Sci. Eng. **28**(1), 1–5 (2012)
8. Huang, Z., et al.: Analysis of the relationship among the surface crack parameters of compacted expansive soil. J. Lanzhou Inst. Technol. **21**(3), 16–20 (2014)

Temperature Spatial and Temporal Distributions of Pavements with Various Paving Materials

Hao Wu[1], Zhichao Zhai[1], Weimin Song[1](✉), and Shu Bai[1,2]

[1] School of Civil Engineering, Central South University, 22 South Shaoshan Rd., Changsha 410075, P.R. China
wsong8@csu.edu.cn

[2] Hunan Provincial Communications Planning, Survey & Design Institute Co., Ltd, 158 North Furong Rd., Changsha, Hunan 410008, P.R. China

Abstract. The mechanical performances and functionalities of pavements are significantly dependent on their service temperatures. It is of great importance to understand the temperature distributions of pavements under various environmental conditions and the influence mechanisms caused by the environmental media, and accuratc and effective estimations of the pavement temperature field in both spatial and temporal dimensions is desired to better predict pavement responses and for a proper pavement design. However, as well known, the temperature distribution in a pavement structure is greatly affected by the paving material and strongly correlated environmental factors, and it is hard to be measured in the field effectively. In this study, numerical simulations based on finite element analysis (FEA) were performed to investigate the temperature distributions of pavements with various paving materials spatially and temporally, and the influences caused by typical environmental media on actual pavements were also discussed based on the simulation results. Four commonly used paving materials, traditional Portland cement concrete (PCC), Portland cement porous concrete (PCPC), dense-graded asphalt concrete (AC), and open-graded friction course (OGFC), were considered for the study. The intensity of solar radiation during daytime and night and wind speed were two environmental factors take into consideration as the environmental media that affect the temperature distributions of the pavement. On the other hand, a series laboratory simulation tests were conducted on the specimens made by various paving materials, and the results from the numerical analysis were compared to those from the laboratory tests in order to verify the effectiveness of the numerical model established in the study. Furthermore, the mechanisms of the environmental media impacting the temperature distributions of the pavements were explored.

Keywords: Pavements with various materials · Temperature distribution · Thermal behaviors · FEA

1 Introduction

Understanding the pavements' behaviors under the effect of various environmental conditions is of great importance nowadays. Asphalt pavement is the main pavement type

S. S. Kim et al. (Eds.): GeoChina 2021, SUCI, pp. 40–51, 2021.
https://doi.org/10.1007/978-3-030-80152-6_4

around the world, which accounts for more than 90% in some countries. For asphalt pavements, the inherent properties of asphalt determine (relatively low albedo) that it has significant endothermic characteristics (Synnefa et al. 2011; Wu et al. 2010; Yinfei et al. 2014). Asphalt pavement has been proved to be a significant contributor to urban heat island effect (Mohajerani et al. 2017). The traditional dense-graded asphalt pavements in a summer climate have a temperature range upwards of 60 °C (Bobes-Jesus et al. 2013; Doulos et al. 2004; Higashiyama et al. 2016). The rutting damage mainly caused by high temperature a typical failure which significantly affects the service. On the other hand, a low temperature makes asphalt stiffer, and the low temperature cracking is easy to occur. The two failure models of asphalt pavements are related to the thermal and viscoelastic behaviors of asphalt (Islam and Tarefder 2015; Song et al. 2018). Therefore, accurate determination of pavement temperature especially surface temperature is essential for pavement analysis and design.

Besides of asphalt pavements, concrete pavement also occupies an important position on a global basis, especially for the heavy-duty transportation (Chung 2012; Mackiewicz and Szydło 2020; Nishizawa et al. 2017). The thermal stress caused by slow seasonal temperature changes and rapid temperature changes are critical factors leading to concrete slab cracking (Westergaard 1927). On the other hand, the stress can also be induced by a uniform change in temperature.

In recent years, open-graded friction course (OGFC) pavements or Portland cement porous concrete (PCPC) pavements are attracting extensive concerns due to the benefits in economy, environment and safety (Peng et al. 2018; Sogbossi et al. 2020; Song et al. 2015, 2017; Stempihar et al. 2012; Wu et al. 2020). OGFC and PCPC pavements follow the philosophy of Low Impact Development (LID). Because of the large porosity, good drainability and noise reduction effectiveness can be obtained. On the other hand, LID philosophy emphasizes the amelioration of urban heat island effect (UHIE). The construction of OGFC or PCPC is an approach to mitigating UHIE (Chen et al. 2019; Ferrari et al. 2020; Wu et al. 2018).

The current research on the pavement temperature field mainly focuses on the laboratory and field tests. Many valuable findings were concluded. Wind speed is one of the most important natural variables on the UHIE (Memon et al. 2010; Morris et al. 2001). Higher wind speeds can help relieve UHIE, promote air circulation, improve cooling systems, and dissipate pollutants (Santamouris 2015). Many measures regarding to pavement structure were proposed to mitigate UHIE, such as the application of cool pavements, reflective pavements, pervious concrete pavements, and porous asphalt pavements (Hassn et al. 2016; Qin 2015; Rossi et al. 2014). Some numerical studies were also conducted to investigate the mechanical or thermal behaviors related to temperatures (Keshavarzi and Kim 2020; Xue et al. 2020; Yang and Chow 2018; Yeon et al. 2013). However, these studies may just consider one factor affecting the thermal-dynamic behaviors of pavements, which cannot fulfil the characterization of pavement performance under the influence of complex factors. On the other hand, thermal behaviors among different pavement types need further examination. A better understanding of the thermal conductivity of various pavements, and the influencing factors, are critical to enhance the pavement performance and the mitigation of UHIE.

2 Objectives and Scope

In this study, finite element analysis (FEA) were performed to investigate the temperature distributions of pavements with various paving materials spatially and temporally. The effect of factors on the thermal behaviors, including intensity of solar radiation and wind speed, were quantitively explored. Four commonly used paving materials, traditional Portland cement concrete (PCC), Portland cement porous concrete (PCPC), dense-graded asphalt concrete (AC), and open-graded friction course (OGFC), were considered for the study. Numerical results were compared with laboratory studies.

3 Finite Element Modeling

3.1 Materials

Materials properties affecting the temperature distribution of pavements include density, radiation absorptivity, specific heat capacity and heat conductivity coefficient (Table 1). The thermal properties of pavement materials and their albedo are quite significant determining factors in their performance and behavior. Albedo values are shown in Table 2, which were measured from laboratory tests conducted in (Wu et al. 2018). Besides, the radiation emissivity is a necessary parameter for calculating temperature distribution. The general value of radiation emissivity for pavement is in range of 0.7 to 0.9. In this study, the value was selected as 0.8.

Table 1. Material properties

	PCC	PCPC	AC	OGFC
Bulk density (kg/m^3)	2569	1956	2324	2000
Radiation absorptivity	0.725	0.875	0.945	0.963
Specific heat capacity (J/kg°C)	1025	904	804	812
Heat conductivity coefficient (W/mK)	1.50	1.28	1.50	1.30

Table 2. Albedo

Specimen	Height of incidence (mm)				Average
	300	400	500	600	
AC	0.048	0.055	0.056	0.054	0.055
OGFC	0.04	0.037	0.036	0.038	0.037
PCC	0.26	0.31	0.27	0.26	0.275
PCPC	0.13	0.12	0.13	0.12	0.125

Table 3 presents the information of pavement structure. Four types of materials were selected as overlay: PCC, PCPC, AC and OGFC.

Table 3. Pavement layers and materials

Layer	Material	Depth (mm)
Overlay	PCC/PCPC/AC/OGFC	60
Underlying layer	AC-20	80
Base	Cement stabilized macadam	200
Subbase	Cement stabilized soil	300
Soil	Sandy soil	1000

3.2 Calculation

The heat exchange effect with the road surface is solar radiation, the convective heat exchange between the air near the ground and the road surface, and the effective radiation of the road surface to the surrounding environment. For asphalt pavements, the related parameters in the following equation can be found in (Herb et al. 2008; Jansson et al. 2006).

$$Q = \alpha_s \cdot Q_s + h(T_a - T_s) + \varepsilon\sigma(T_{a,K} - T_{s,K})$$

Q: total heat flux into the pavement structure (W·m^{-2})
α_s: absorption rate of road surface to solar radiation
Q_s: Total solar radiation density (W·m^{-2})
h: Convection heat transfer coefficient (W·m^{-2}·K^{-1})
ε : Radiation emissivity of pavement structure
σ: Steven-Boltzmann constant, 5.67 × 10^{-8} (W·m^{-2}·K^{-4})
T_a,T_s: air and road surface temperatures (°C)
$T_{a,K}$,$T_{s,K}$: Kelvin temperatures of air and road surface (K).

Inside of the pavement, there is a heat conduction process, which makes the heat flow between the pavement layers. Transient heat transfer, when only conduction is considered, can be expressed as (Cengel 2003):

$$\frac{\partial T}{\partial t} - \alpha\left(\frac{\partial^2 T}{\partial x^2} + \frac{\partial^2 T}{\partial y^2} + \frac{\partial^2 T}{\partial z^2}\right) = 0$$

T : temperature field function with respect to time and space (°C)
α: thermal diffusivity (m^2·s^{-1}).

In the calculation of the solar radiation, the value is regarded as before sunrise and after sunset. Solar radiation during the daytime has a sinusoidal distribution:

$$q = -\frac{S}{t_2 - t_1}\cos\left[\frac{\pi}{t_2 - t_1}(2t + t_2 - t_1 - 2t_m)\right] + \frac{S}{t_2 - t_1}$$

q : radiation intensity (J/(m^2·h))
t: time (h)
S: total daily radiation (MJ/m^2)

t_1,t_2: sunrise and sunset time

t_m: peak radiation time.

In Changsha, China, the solar radiation was obtained based on meteorological information, shown in Table 4. The radiation around a whole year was presented.

Table 4. Meteorological information of solar radiation in Changsha area

	Total daily radiation (MJ/m^2)	Sunrise time	Sunset time	Peak radiation time
Spring (April)	9.68	6:00	19:00	12:30
Summer (June)	19.69	5:30	19:30	12:30
Winter (January)	4.45	7:20	18:00	12:40

Air convection near the ground is an important factor affecting the temperature field of the road surface (Table 5).

$$T = \frac{T_{\max} - T_{\min}}{2} \cos\left[\frac{\pi(t + 24 - t_{\max})}{t_{\min} - t_{\max} + 24}\right] + \frac{T_{\max} + T_{\min}}{2}, \quad 0 \le t \le t_{\min}$$

$$T = -\frac{T_{\max} - T_{\min}}{2} \cos\left[\frac{\pi(t - t_{\min})}{t_{\max} - t_{\min}}\right] + \frac{T_{\max} + T_{\min}}{2}, \quad t_{\min} \le t \le t_{\max}$$

$$T = \frac{T_{\max} - T_{\min}}{2} \cos\left[\frac{\pi(t - t_{\max})}{t_{\min} - t_{\max} + 24}\right] + \frac{T_{\max} + T_{\min}}{2}, \quad t_{\max} \le t \le 24$$

t : time (h)T: temperature (°C)$T_{\min}$, $T_{\max}$: daily lowest temperature, daily highest temperature (°C)$t_{\min}$, t_{maxn}: time of the lowest and highest daily temperature.

Table 5. Temperature information in Changsha area

	Daily lowest temperature (°C)	Time of the lowest temperature	Daily highest temperature (°C)	Time of the lowest temperature
Spring (April)	9.68	6:00	19:00	12:30
Summer (June)	19.69	5:30	19:30	12:30
Winter (January)	4.45	7:20	18:00	12:40

Wind could affect the pavement surface temperature by affecting the convective heat transfer (Postgård and Lindqvist 2001). The relationship between wind speed and convective heat transfer was shown in Table 6.

In order to simplify the analysis, heat transfer by conduction, solar radiation and wind speed effect were considered in the overlay while heat transfer by conduction was considered in underlying layer, base, subbase and soil.

Table 6. The relationship between wind speed and convective heat transfer coefficient

Wind speed (m/s)	Convection heat transfer coefficient ($W \cdot m^{-2} \cdot K^{-1}$)
0	6
1	10
2	14
3	18
4	22
5	25
6	28
7	32

3.3 Model Establishment

The length of the model was 24 m, and the width and height were 6 and 1.64 respectively. DC3D20 unit was selected. The model is shown in Fig. 1. The effect of the bonding materials between pavement layers was neglected in this study. An assumption was made that the bonding between the layers is good, and the heat conduction between the layers is continuous.

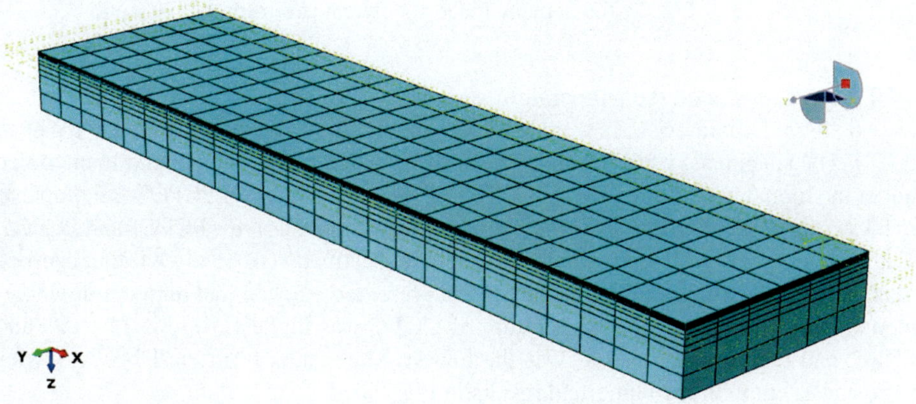

Fig. 1. Model

4 Results and Discussion

4.1 Temperature Distributions of Different Pavements

Temperature distributions of difference pavements in winter season were examined. Relationships between time and temperature were plotted in Fig. 2. For all pavement types, temperature distributions in the vertical direction were the same, e.g., temperatures gradually decreased along with the depth increase. Considering the temperature on the top of overlays, temperature of OGFC was the largest, followed by AC and PCPC, and temperature of PCC was the lowest. Temperatures asphalt pavements (OGFC and AC) were larger than those of concrete pavements (PCC and PCPC), which was due to the fact that asphalt material has a higher radiation absorption rate and a stronger ability to absorb solar radiation. Temperatures of OGFC and PCPC pavements, which were high porosity materials, were larger than those of AC and PCC. The results were ascribed to that the significant void structures induced higher radiation absorption rates.

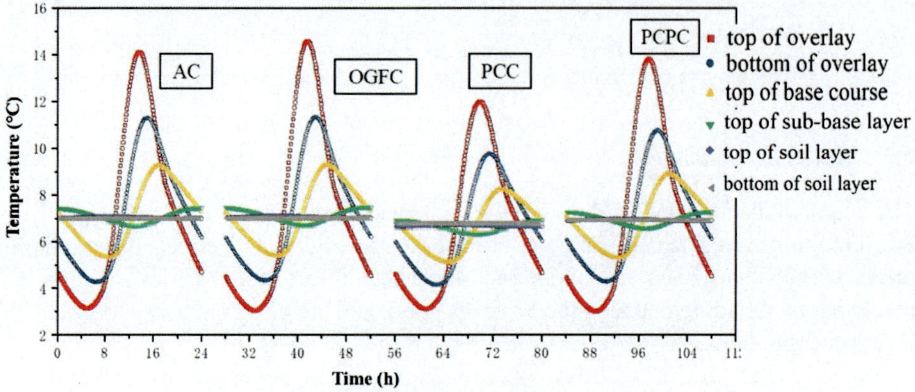

Fig. 2. Temperature fields of different pavements

Figure 3 illustrated the temperature variations of some specimens from laboratory investigation. Laboratory tests were conducted according to the procedures in (Wu et al. 2018). The specimens were irradiated with the infrared lamps for 5 h and then cooled under the room temperature (around 20 °C) for 12 h. Two types of PCPC were prepared in these tests. The difference was mainly in aggregate gradation for PCPC1 and PCPC2. With relatively lower albedo values, the asphalt specimens (AC and OGFC) absorbed much more heat from the thermal sources and presented much higher temperature increments. During the first 5 h, temperature of OGFC was highest, followed by AC and PCPC, and temperature of PCC was the lowest. The results from the laboratory tests were consistent with the numerical results in Fig. 2.

Figure 4 shows the temperature change rates with time for different pavement materials. The rate of change is the differential of temperature with respect to time in Fig. 2. The temperature change rate of the four pavement types shows similar changes under the same weather conditions. The peak rates of the overlay surface appeared at 10 AM, indicating the temperatures at this time obtained the fastest increasing rate. The temperature change rates subsequently decreased, and rates approached to 0 at 1 PM, indicating temperatures at the top of overlays got the peak values.

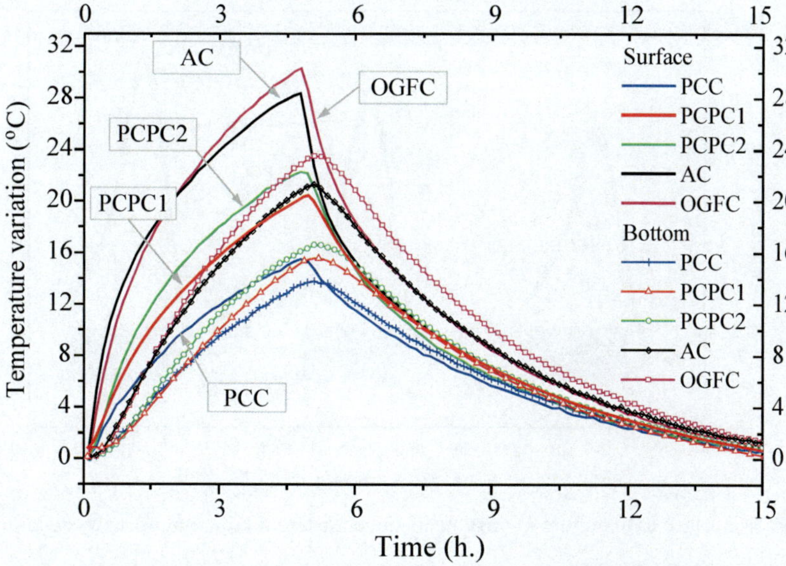

Fig. 3. Surface and bottom temperatures of different materials

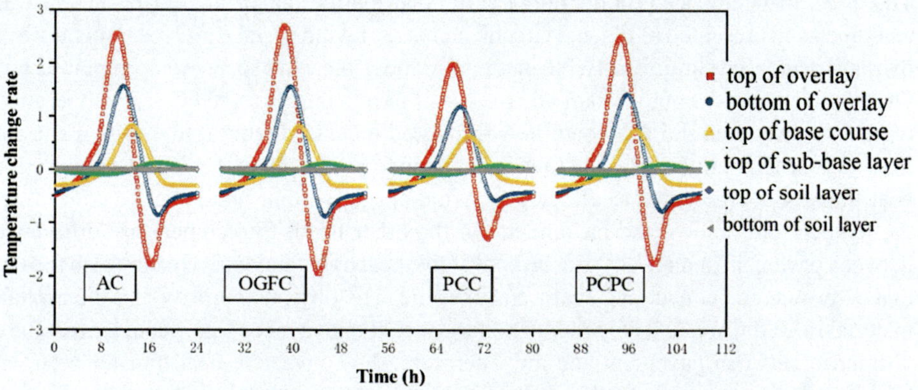

Fig. 4. Temperature change rates of different pavements

4.2 Temperature Distributions Under Different Radiation Conditions

Figure 5 shows the response of the temperature field of the four types surface materials under different radiation intensities. Under the same radiation intensity value, the temperatures of asphalt pavements (OGFC and AC) were higher than those of concrete pavements (PCPC and PCC), and the temperatures of open-graded materials were correspondingly higher than those of dense-graded materials. This difference gradually increases as the radiation intensity increases. When the radiation intensity increases from 0.8 times to 1.2 times, the maximum temperature of the OGFC increased by about 3 °C, while the maximum temperature of the PCC increased by about 2 °C.

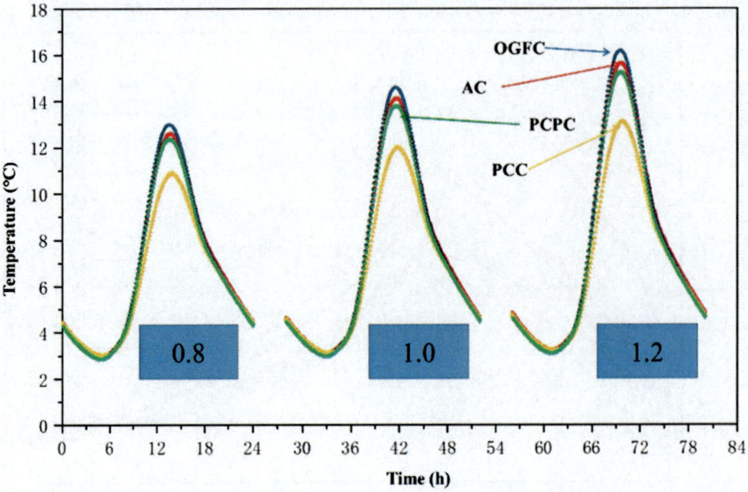

Fig. 5. Surface temperature of pavements under different radiation intensity conditions

4.3 Temperature Distributions Under Different Wind Speeds

The effect of wind speed on the temperature distribution on surfaces of pavement was examined and presented in Fig. 6. With the increase of wind speed, surface temperature of overlays decreased significantly. No matter how much the wind speed was, temperature of OGFC pavement was higher than other types of pavements. For OGFC, the temperature reduction was about 4.5 °C when the wind speed increased from 0 to 4 m/s; while for PCC, whose temperature was the lowest, the reduction was about 2.5 °C. The convective heat transfer coefficients are affected by wind speed, so that wind speed could affect the temperature field of the pavement. On the other hand, the temperature difference between pavement and the air, and the convective heat transfer coefficient, are also factors that determine the convective heat transfer value. The obvious porosity and the asphalt material in OGFC make it absorb more energy, leading to a more significant temperature difference between pavement and air. Therefore, the convective heat transfer between OGFC and air is more likely affected; so, the temperature reduction is more obvious for OGFC.

Laboratory tests were also performed to investigate the effect of wind on temperature change of these materials. The results can be observed in Fig. 7. Similar with Fig. 7, compared to the temperatures of specimens under windless conditions, lower temperatures were observed on all the specimens and generally the greater the speed of wind, the larger reduction of the temperatures. Some differences were also obtained that temperatures of AC samples were highest, which was not consistent with the results in Fig. 6. More tests will be conducted in future to check the difference between laboratory and numerical results.

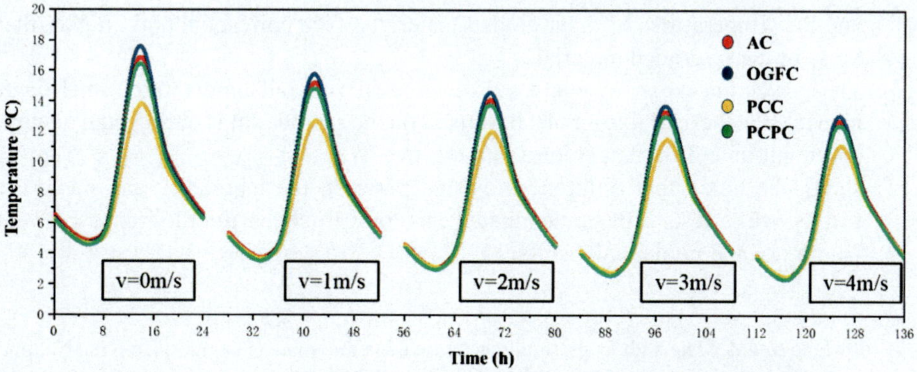

Fig. 6. Surface temperature of different pavement under different wind speed

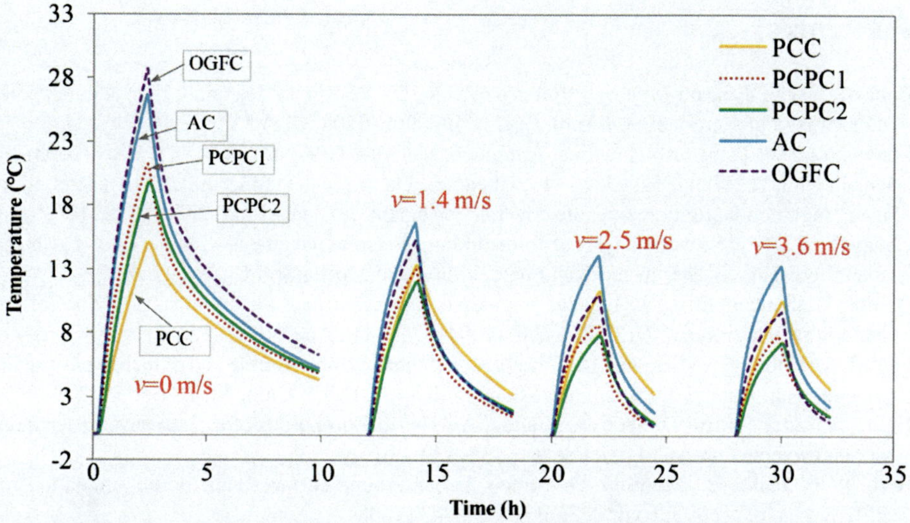

Fig. 7. Temperatures of different specimens under the effect of wind

5 Conclusions

Finite element analysis was conducted to performed to investigate the temperature distributions of pavements with various paving materials spatially and temporally, and the influences caused by typical environmental media on actual pavements were also discussed based on the simulation results. Laboratory results were also compared with numerical results.

(1) Due to the relatively lower albedo values, asphalt pavements usually absorb more radiation and exhibit higher temperatures than those of the cement concrete pavements during daytime.
(2) Under the same radiation intensity value, the temperatures of asphalt pavements (OGFC and AC) were higher than those of concrete pavements (PCPC and PCC),

and the temperatures of open-graded materials were correspondingly higher than those of dense-graded materials.

(3) The convection caused by wind could decrease the overall temperature, which could alleviate the thermal impact of a pavement on the surroundings during both daytime and nighttime. The temperature ranking was $T_{OGFC} > T_{AC} > T_{PCPC} > T_{PCC}$. It should be noted that differences existed between the laboratory and numerical results. More tests will be conducted in future to check the difference between laboratory and numerical results.

Acknowledgement. The authors gratefully acknowledge the financial support from the National Natural Science Foundation of China (Grant No.: 51778638).

References

Bobes-Jesus, V., Pascual-Muñoz, P., Castro-Fresno, D., Rodriguez-Hernandez, J.: Asphalt solar collectors: a literature review. Appl. Energy **102**, 962–970 (2013)

Cengel, Y.: Heat Transfer: A Practical Approach. McGraw-Hill, New York, NY, USA (2003)

Chen, J., Chu, R., Wang, H., Zhang, L., Chen, X., Du, Y.: Alleviating urban heat island effect using high-conductivity permeable concrete pavement. J. Clean. Prod. **237**, 117722 (2019)

Chung, Y.: Thermal Stress Analysis of Jointed Plain Concrete Pavements Containing Fly Ash and Slag. Louisiana State University and Agricultural and Mechanical College (2012)

Doulos, L., Santamouris, M., Livada, I.: Passive cooling of outdoor urban spaces. The role of materials. Solar Energy **77**(2), 231–249 (2004)

Ferrari, A., Kubilay, A., Derome, D., Carmeliet, J.: The use of permeable and reflective pavements as a potential strategy for urban heat island mitigation. Urban Clim. **31**, 100534 (2020)

Hassn, A., Chiarelli, A., Dawson, A., Garcia, A.: Thermal properties of asphalt pavements under dry and wet conditions. Mater. Des. **91**, 432–439 (2016)

Herb, W.R., Janke, B., Mohseni, O., Stefan, H.G.: Ground surface temperature simulation for different land covers. J. Hydrol. **356**(3), 327–343 (2008)

Higashiyama, H., Sano, M., Nakanishi, F., Takahashi, O., Tsukuma, S.: Field measurements of road surface temperature of several asphalt pavements with temperature rise reducing function. Case Stud. Constr. Mater. **4**, 73–80 (2016)

Islam, M.R., Tarefder, R.A.: Coefficients of thermal contraction and expansion of asphalt concrete in the laboratory. J. Mater. Civil Eng. **27**(11), 04015020 (2015)

Jansson, C., Almkvist, E., Jansson, P.-E.: Heat balance of an asphalt surface: observations and physically-based simulations. Meteorol. Appl. **13**(2), 203–212 (2006)

Keshavarzi, B., Kim, Y.R.: A dissipated pseudo strain energy-based failure criterion for thermal cracking and its verification using thermal stress restrained specimen tests. Constr. Build. Mater. **233**, 117199 (2020)

Mackiewicz, P., Szydło, A.: Thermal stress analysis in concrete pavements. J. Transport. Eng. **146**(3), 06020002 (2020)

Memon, R.A., Leung, D.Y.C., Liu, C.-H.: Effects of building aspect ratio and wind speed on air temperatures in urban-like street canyons. Build. Environ. **45**(1), 176–188 (2010)

Mohajerani, A., Bakaric, J., Jeffrey-Bailey, T.: The urban heat island effect, its causes, and mitigation, with reference to the thermal properties of asphalt concrete. J. Environ. Manage. **197**, 522–538 (2017)

Morris, C.J.G., Simmonds, I., Plummer, N.: Quantification of the influences of wind and cloud on the nocturnal urban heat island of a large city. J. Appl. Meteorol. Climatol. **40**(2), 16–182 (2001)

Nishizawa, T., Koyanagawa, M., Takeuchi, Y., Kubo, K., Yoshimoto, T.: Thermal stress calculation method for concrete pavement based on temperature prediction and finite element method analysis. Transport Res. Rec. **2640**(1), 104–114 (2017)

Peng, H., Yin, J., Song, W.: Mechanical and hydraulic behaviors of eco-friendly pervious concrete incorporating fly ash and blast furnace slag. Appl. Sci. **8**(6), 859 (2018)

Postgård, U., Lindqvist, S.: Air and road surface temperature variations during weather change. Meteorol. Appl. **8**(1), 71–83 (2001)

Qin, Y.: A review on the development of cool pavements to mitigate urban heat island effect. Renew. Sustain. Energy Rev. **52**, 445–459 (2015)

Rossi, F., Pisello, A.L., Nicolini, A., Filipponi, M., Palombo, M.: Analysis of retro-reflective surfaces for urban heat island mitigation: a new analytical model. Appl. Energy **114**, 621–631 (2014)

Santamouris, M.: Analyzing the heat island magnitude and characteristics in one hundred Asian and Australian cities and regions. Sci. Total Environ. **512–513**, 582–598 (2015)

Sogbossi, H., Verdier, J., Multon, S.: Permeability and damage of partially saturated concrete exposed to elevated temperature. Cement Concr. Compos. **109**, 103563 (2020)

Song, W., Huang, B., Shu, X.: Influence of warm-mix asphalt technology and rejuvenator on performance of asphalt mixtures containing 50% reclaimed asphalt pavement. J. Clean. Prod. **192**, 191–198 (2018)

Song, W., Shu, X., Huang, B., Woods, M.: Factors affecting shear strength between open-graded friction course and underlying layer. Constr. Build. Mater. **101**, 527–535 (2015)

Song, W., Shu, X., Huang, B., Woods, M.: Influence of interface characteristics on the shear performance between open-graded friction course and underlying layer. J. Mater. Civ. Eng. **29**(8), 04017077 (2017)

Stempihar, J.J., Pourshams-Manzouri, T., Kaloush, K.E., Rodezno, M.C.: Porous asphalt pavement temperature effects for urban heat island analysis. Transport Res. Rec. **2293**(1), 123–130 (2012)

Synnefa, A., Karlessi, T., Gaitani, N., Santamouris, M., Assimakopoulos, D.N., Papakatsikas, C.: Experimental testing of cool colored thin layer asphalt and estimation of its potential to improve the urban microclimate. Build. Environ. **46**(1), 38–44 (2011)

Westergaard, H.M.: Analysis of stresses in concrete due to variations of temperature, 6th Annual, pp. 201–215. Meeting Highway Research Board. National Research Council, Washington, DC (1927)

Wu, H., Sun, B., Li, Z., Yu, J.: Characterizing thermal behaviors of various pavement materials and their thermal impacts on ambient environment. J. Clean. Prod. **172**, 1358–1367 (2018)

Wu, H., Yu, J., Song, W., Zou, J., Song, Q., Zhou, L.: A critical state-of-the-art review of durability and functionality of open-graded friction course mixtures. Constr. Build. Mater. **237**, 117759 (2020)

Wu, Q., Zhang, Z., Liu, Y.: Long-term thermal effect of asphalt pavement on permafrost under an embankment. Cold Reg. Sci. Technol. **60**(3), 221–229 (2010)

Xue, W., Shen, L., Jing, W., Li, H.: Permeability evolution and mechanism of thermally damaged basalt fiber-reinforced concrete under effective stress. Constr. Build. Mater. **251**, 119077 (2020)

Yang, Y., Chow, C.L.: Transient temperature fields and thermal stress fields in glazing of different thicknesses exposed to heat radiation. Constr. Build. Mater. **193**, 589–603 (2018)

Yeon, J.H., Choi, S., Won, M.C.: In situ measurement of coefficient of thermal expansion in hardening concrete and its effect on thermal stress development. Constr. Build. Mater. **38**, 306–315 (2013)

Yinfei, D., Qin, S., Shengyue, W.: Highly oriented heat-induced structure of asphalt pavement for reducing pavement temperature. Energy Build. **85**, 23–31 (2014)

Used Paper Fibers for Sustainably Enhancing the MICP Stabilization of Sand

Meiqi Chen[1](✉), Sivakumar Gowthaman[2], Kazunori Nakashima[2], Shin Komatsu[3], and Satoru Kawasaki[2]

[1] Graduate School of Engineering, Hokkaido University, Sapporo, Japan
[2] Faculty of Engineering, Hokkaido University, Sapporo, Japan
[3] Meiwa Seishi Genryo Co., Ltd., Osaka, Japan

Abstract. The increasing awareness of the energy crisis and environmental protection has led to a proliferation of studies on novel ground improvement techniques. One of these techniques is microbially induced carbonate precipitation (MICP), which sustainably applies the microorganisms for soil stabilization purposes. During the process, calcium carbonate is achieved bio-chemically within the embedded soil, enhancing the strength and stiffness. For several decades, fiber materials have been a prime part of soil improvement on account of its desirable mechanical characteristics. Therefore, this work aims to introduce the fiber onto the MICP, demonstrating how the fiber material could enhance the MICP process in the sand. Used paper fibers, one type of recycled material, were chosen herein as an environmental-friendly option, and a series of experiments were conducted on the sand with different ratios (1–8%) of fiber. The findings suggest that the fiber addition enhanced the immobilization of the bacteria cells and provided favourable conditions for bacterial performance and survival. The calcium carbonate measurements revealed that the fiber addition could significantly yield the precipitation content, increasing the unconfined compressive strength (UCS). However, the optimum fiber content corresponding to the highest UCS was found to be 1%, and further addition appeared to suppress the UCS of sand. Overall, the study has demonstrated that the used paper fiber could be effectively reused in the MICP technique, enabling the pathway with several desirable merits.

1 Introduction

Recent trends in sustainable development have heightened the need for the introduction of new technologies into ground improvement techniques (DeJong et al., 2010). Microbially induced carbonate precipitation (MICP) is relatively a new environmentally friendly method, which has rapidly gained greater attention as a green technology for the future (DeJong et al., 2010). The process of MICP highly relies on metabolic activity and a series of biochemical reactions, persuading the calcium carbonate cement within the embedded soil matrix. Urease produced by soil bacteria (referred to as ureolytic bacteria) catalyzes the hydrolysis of urea (Eq. (1)), hence favorably increase the pH of the reaction medium. In the presence of calcium ions, calcium carbonate precipitates in nucleation sites provided by the bacterial cells (Eq. (2)), which binds the soil particles and acts in a

S. S. Kim et al. (Eds.): GeoChina 2021, SUCI, pp. 52–64, 2021.
https://doi.org/10.1007/978-3-030-80152-6_5

way similar to that of bridges among soil particles (Ivanov and Chu 2008; Dejong et al., 2013).

$$CO(NH_2)_2 + 2H_2O \xrightarrow{Urease} 2NH_4^+ + CO_3^{2-} \tag{1}$$

$$CO_3^{2-} + Ca^{2+} \overset{Bacterial\ cell}{\rightarrow} Cell - CaCO_3 \downarrow \tag{2}$$

To date, many research works have been focused on MICP technique from several perspectives, addressing on (i) biochemical factors affecting MICP (Naveed et al., 2020), (ii) mechanical properties of MICP treated soils (Feng and Montoya, 2016), (iii) durability properties under various environmental factors (Gowthaman et al., 2020; Liu et al., 2019) and (iv) scaling-up and associated challenges (van Paassen et al., 2010). In spite of the considerable interests and achievements in MICP, some problems still remain unsolved. One of such main obstacles is the undesirable brittle failure of MICP treated soil with high post-peak loss (Fang et al., 2020).

It has been reported that the fiber material had played a significant role in soil stabilization since 1969 (Gowthaman et al., 2018). Particularly, fiber additions have often been reported to be declining the brittle characteristics, instead, preferably promoting ductility of the material (Park, 2011; Imran et al., 2020). For instance, fiber was used to reinforce the soil-cement mixture, reveled a remarkable enhancement in both mechanical and durability behaviors (Khattak and Alrashidi, 2006). As mentioned before, due to the advantages of fiber addition, a diverse range of synthetic fibers and natural fibers are produced and broadly applied in the Geotechnical field. In this research, the used paper fiber, one mixture of waste materials collected from a paper mill, has been studied on its impact on MICP-treated sands. In fact, utilizing the waste material in the engineering application is always encouraged, as it perfectly meets the sustainable development requirements. Moreover, it is assumed that the main composition (cellulose) of the paper fiber aids to enhance the bacterial survivability in the sand during MICP. As the available information regarding fiber-microbial interactions is limited, this study has been potentially undertaken to deepen understanding of the fiber-MICP pathway.

This study seeks to examine the feasibility of introducing used paper fibers onto MICP technology and figure out whether these fibers could improve the efficiency of MICP or not. For this purpose, three sets of experiments were conducted to ascertain the function of used paper fibers in treated samples. Several parameters were discussed herein, including the bacterial retention rate, estimated UCS, and carbonate content of different parts of treated samples. For the micro-scale observation, a scanning electron microscope (SEM) and fluorescence microscope were used to identify the microstructure and the live bacteria in samples, respectively.

2 Materials and Methods

2.1 Tested Sand

The sand used in this study is one of the standard laboratory sands in Japan: Mikawa sand (No. 4), with a mean particle size of 0.87 mm. The grain size distribution of Mikawa sand is presented in Fig. 1. Table 1 shows the intrinsic properties and compositions of Mikawa sand by using a dispersive X-Ray Fluorescence (XRF) spectrometer (JSX-3100R II JOEL, Japan).

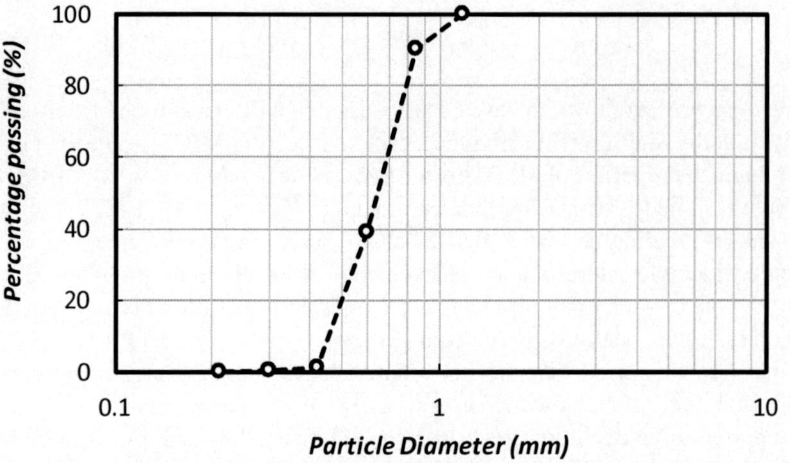

Fig. 1. Grain size distribution of Mikawa sand

Table 1. The fundamental properties and compositions of Mikawa sand

Basic properties		Composition (%) based on XRF analysis							
Moisture Content, %	pH	MgO	Al_2O_3	SiO_2	K_2O	CaO	TiO_2	Fe_2O_3	Na_2O
0	7.01	0.11	0.89	97.65	0.18	0.03	0.04	0.43	0.04

2.2 Used Paper Fiber

The fibers used in this study were the waste material, obtained from the plant of Meiwa Seishi Genryo Company, Japan. The fiber matrix was used without any pre-treatments, and that was found to consist of different types of fibers, along with the dust produced during the recycling process. The moisture content of fibers was around 3.6%. To obtain an estimated size of fibers, SEM (scanning electron microscopy) analysis was conducted; the physical appearance and the SEM image of the fibers are presented in Fig. 2(a) and 2(b), respectively. The length of the fibers was found in the range between 20–500 μm.

Fig. 2. (a) Physical appearance and (b) SEM image of the fiber material

2.3 Bacteria and Cementation Media

The ureolytic bacteria used in this study was *Lysinibacillus xylanilyticus*, a gram positive, rod shape species (8–10 μm long), which was previously identified from Hokkaido, Japan. The urease activity of *Lysinibacillus xylanilyticus* was found to be high at a mild temperature; the activity at 25 °C peaked around 3.5 U/mL after 72 h cultivation (Gowthaman et al., 2019a). Therefore, in this study, the temperature during cultivation and curing period was set to be 25 °C. Further detail on the isolation and identification process of the bacteria can be found from the previous report (Gowthaman et al., 2019b). The culture medium used to cultivate the bacteria and the cementation media used for the MICP treatment are listed clearly in Table 2.

Table 2. Ingredients of culture and cementation solutions

Solution	Substance	Amount (g/L)
NH_4-YE medium	Tris-buffer	15.75
	Ammonium sulfate	10
	Yeast extract	20
Cementation media	$CaCl_2$	55.5
	Urea	30
	Nutrient broth	1.5

2.4 Preparation and experiment design

During the preparation, 60 g of Mikawa sand was mixed well with fibers (until the mixture appears homogeneous) and filled into the vertically positioned syringe (30 mm in diameter, 50 mL in volume). A number of samples were prepared with different fiber in the range between 1–8% by sand weight, and the range was chosen to be consistent with the previous research works (Gowthaman et al., 2018). The bacteria were cultivated under shaking incubation at 25 °C and 160 rpm, which was proved to be the optimum temperature for *Lysinibacillus xylanilyticus*. When the population (OD_{600}) of the bacteria reached its maximum (after around 72 h), they were harvested (at the OD_{600} of around 3.5 ~ 4) for the treatment.

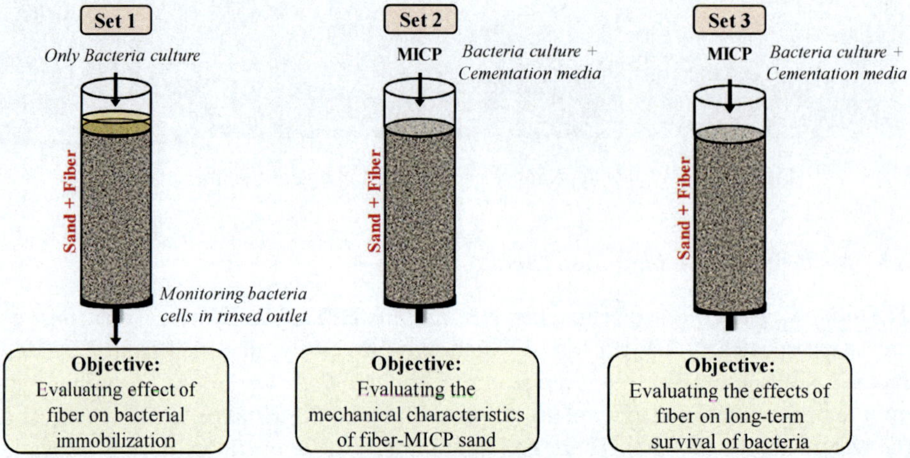

Fig. 3. Schematic outline of this study

2.4.1 Experiment: Set 1

To study the impact of fibers on bacterial performance, immobilization tests were performed (Set 1, Fig. 3). To the samples prepared with different fiber content, 25 mL of bacterial solution was injected on the surface of the syringe columns and remained undisturbed for one hour. Subsequently, the same volume of distilled water was injected to rinse out the bacteria solution. During the rinsing stage, the outlet valve was opened, and 5 mL of outflow solution was collected per rinse. Five numbers of rinsing were done by repeating this procedure. The optical density (OD_{600}) of collected solutions was examined by spectrophotometry to evaluate the bacterial retention.

2.4.2 Experiment: Set 2

Set 2 (Fig. 3) was to establish an understanding on how the mechanical properties are influenced by different fiber additions. MICP treatment was performed on the specimens placed in the incubator of 25 °C. At first, 12 mL of bacteria culture was injected at the specimen top and remained saturated for 2 h, followed by the injection of cementation solution (same volume as bacteria culture). During the two weeks of treatment, the cementation solution was injected every 24 h, and the bacteria was once again injected on the 8th day of the treatment. The details of the test cases are summarized in Table 3.

Table 3. Detail of the test cases (Experiment: Set 2)

Case	Fiber content, % by sand weight	Bacteria solution		Cementation solution		Treatment Duration
		Amount	Injection	Amount	Injection	
1	0%	12 mL (OD_{600} = 4)	Once a week	12 mL (0.5 mol/L)	Every 24 h	2 weeks
2	1%					
3	2%					
4	4%					
5	8%					

Table 4. Detail of the test cases (Experiment: Set 3)

Type	Case	Fluorescence microscopy analysis
Mikawa sand only	A-1	Day-3
	A-2	Day-7
	A-3	Day-14
Mikawa sand + 2% fiber addition	B-1	Day-3
	B-2	Day-7
	B-3	Day-14

2.4.3 Experiment: Set 3

As the natural fiber consists of energy sources of bacteria (Teng et al., 2020), a reasonable expectation is that fiber addition could possibly enhance bacterial survivability. In order to prove the hypothesis, Set 3 experiment (Fig. 3) was carried out basically in the same way as Set 2, aiming to find out the effect of fibers on bacterial survivability in a two-week treatment. Due to the discrepancy of live and dead bacteria on their membrane integrity, when stained by chemicals and excited by ultraviolet radiation, the green color would show on live cells, whereas dead cells would be in red. This protocol was used to study the survivability using fluorescence microscopy. For that, two sets of samples were treated (Set 3) in the same way as Set 2. However, in order to better observe the survivability of bacteria in different samples, the bacteria solution was injected only at the beginning of the treatment. The samples were evaluated at different treatment durations (Day 3, 7, and 14), and the test cases are presented in Table 4.

2.5 Evaluation of Treated Specimens

After the treatment, specimens were oven-dried at 60 °C for 48 h, followed by the estimation of UCS (unconfined compressive strength) by needle penetration tests. Needle penetration test is an ISRM (International Society for Rock Mechanics) recommended method and is widely used to estimate the strength of the soft rocks and cemented soil specimens. During the measurement, the penetration depth of the needle and the penetration resistance are recorded. The regression relationship (Eq. 3) that has been developed by analyzing 114 natural soft rock samples and 50 cemented soil samples was used for the estimation (Amarakoon and Kawasaki 2018; Danjo and Kawasaki 2016).

$$log\ (y) = 0.978\ log\ (x) + 2.621 \tag{3}$$

where y is the UCS; x is the penetration gradient (N/mm) determined by the penetration depth and penetration resistance.

The calcium carbonate content of treated samples was measured using a pressure gauge equipped with an airtight container, which is a simplified method to estimate the carbonate content, developed by Fukue et al. (1999). Firstly, 2 g of dry samples and 20 mL of HCl solution (2 mol/L) were placed into the container (without contact). Then, the container was sealed and equipped with a pressure gauge. Finally, the reaction was triggered by shaking the device to mix the samples and HCl solution. When the pressure change is negligible, the value was recorded. By using the calibration curve, the carbonate content was obtained in percentage as the ratio between the weight of carbonate and the weight of sand before the treatment.

3 Results and Discussion

3.1 Effect of Fiber Addition on Bacterial Response

Figure 4 presents the results of the bacterial immobilization test. It can be seen from Fig. 4(a) that the bacteria cells are tended to be washed out in all the cases during rinsing; however, the bacteria population at the outlet decreases with the increase in the fiber content. From the results, the bacteria retention rate was estimated as per the following equation (Eq. (4)).

$$\text{Bacteria retention rate } (\%) = 1 - \frac{\text{OD}_{600} \text{ of the outlet}}{\text{OD}_{600} \text{ of the injection}} \tag{4}$$

Figure 4(b) demonstrates a remarkable increase in bacterial retention rate when the fiber addition is increased. It is worth noting that the bacterial retention rate is found to be low (around 11%) in the control specimen (without fiber). On the other hand, when the fiber addition was 2%, around 57% of bacteria cells were immobilized in the sample, and only a small loss (less than 25%) of bacteria was observed in samples with the fiber addition of more than 4%. The results reported herein show a very good agreement with the previous work by Zhao et al. (2020), in which the 2% addition of activated carbon-fiber felt was reported to enhance the retention rate up to 54%.

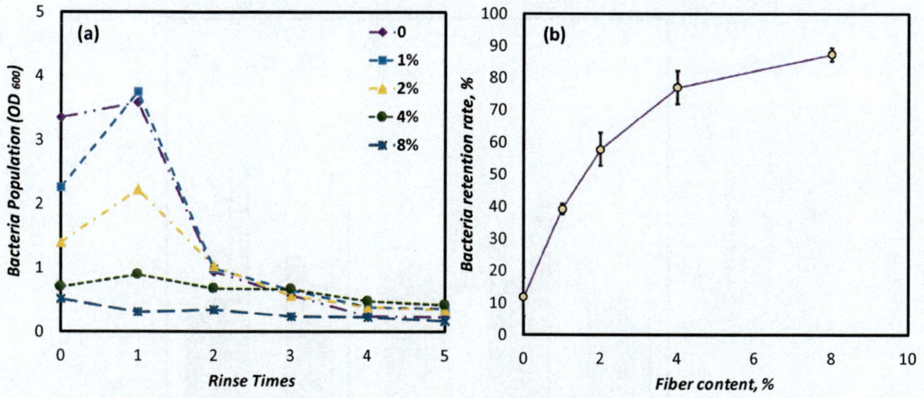

Fig. 4. Bacterial immobilization: (a) rinsing test and (b) retention rate with varying fiber content

3.2 Mechanical Characteristics of Fiber-MICP Sand

Figure 5 presents the UCS results obtained from needle penetration tests for the MICP sand with varying fiber contents. It can be seen that the estimated UCS increased significantly with a small fiber addition of 1%, whereas the strength came out to be low when the fiber addition was increased to 2%. Closer inspection of Fig. 5 further shows that the addition of fiber materials changed the strengthening pattern in specimens. For instance, in the 0% fiber case, higher strength is achieved at a deeper zone (i.e., bottom of the specimen); however, in the 8% fiber case, the top part is the most solidified while the rest remained less cemented. This reveals that the presence of fiber had a prominent role in the spatial distribution of bacteria cells during the injection. Nevertheless, from the observation, 1% fiber content is likely to be the optimum for better MICP response.

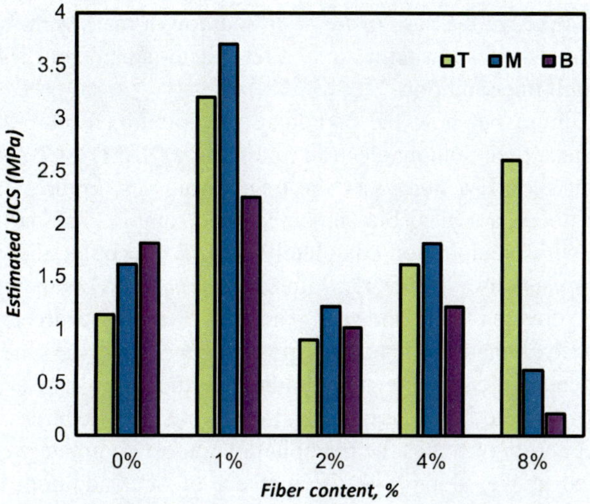

Fig. 5. Estimated UCS at the top (T), middle (M), and bottom (B) parts of the specimens

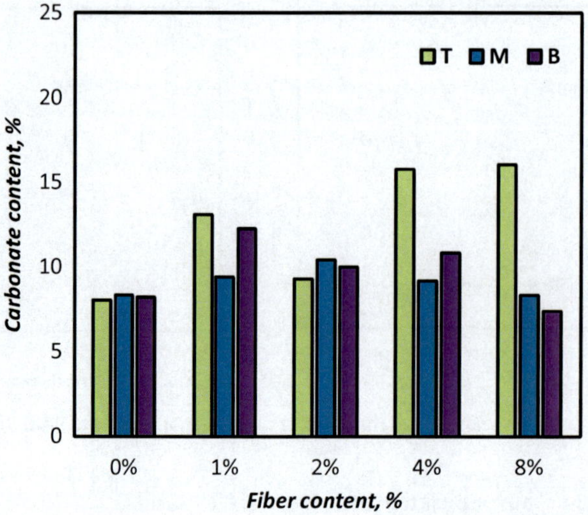

Fig. 6. Carbonate content at top (T), middle (M), and bottom (B) parts of the specimens

Several factors are known to be responsible for the effectiveness of MICP, among which the carbonate content is one of the most critical parameters that must be taken into consideration (Gowthaman et al., 2020). The results obtained from the analysis of carbonate content are compared in Fig. 6. What is striking here is that the carbonate precipitation at the top part was 1.5 times more than the average when the fiber addition was up to 4%. In cases with fiber addition, fiber materials filled the pore spaces between the sand particles and worked as a filter when the bacteria solution seeping through the sand column. This explanation is supported by the results of the bacterial immobilization test in Fig. 4, that the higher fiber addition contains in the sample, the more bacteria remain in the sample. In the case of no fiber addition, a massive loss of bacteria was observed after the first rinse, whereas only a few bacteria came out with the outflow in the cases with high fiber addition.

In order to figure out how the carbonate precipitation occurred in the samples with fiber materials, the scanning electron microscope (SEM) analysis was conducted. Figure 7 presents the SEM images of the treated samples. Figure 7(a) illustrates the sample without fibers, a typical bio-cemented sand sample. The particle surface was almost covered with precipitation completely, and in the contact between particles, a carbonate bridge was observed clearly. Figure 7(b–e) can be compared with Fig. 7(a). The precipitation covered fibers and sand particles to varying degrees. In the case with 1% fiber addition, carbonate precipitations mostly happened on the sand particles. As the fiber addition increases, carbonate precipitations on the fibers appear to be increasing. notably, in the 8% fiber addition case, soil particles were likely to be covered with a small amount of precipitation, whereas the precipitation almost completely coated fibers.

As mentioned above, along with the increase of fiber addition, the less strength was gained. Combined with the micro-scale images above, the answer is possibly the reduction of the effective bonding between sand particles. In other words, the bonding between fibers, or fibers, and sand particles is weaker than that between sand particles.

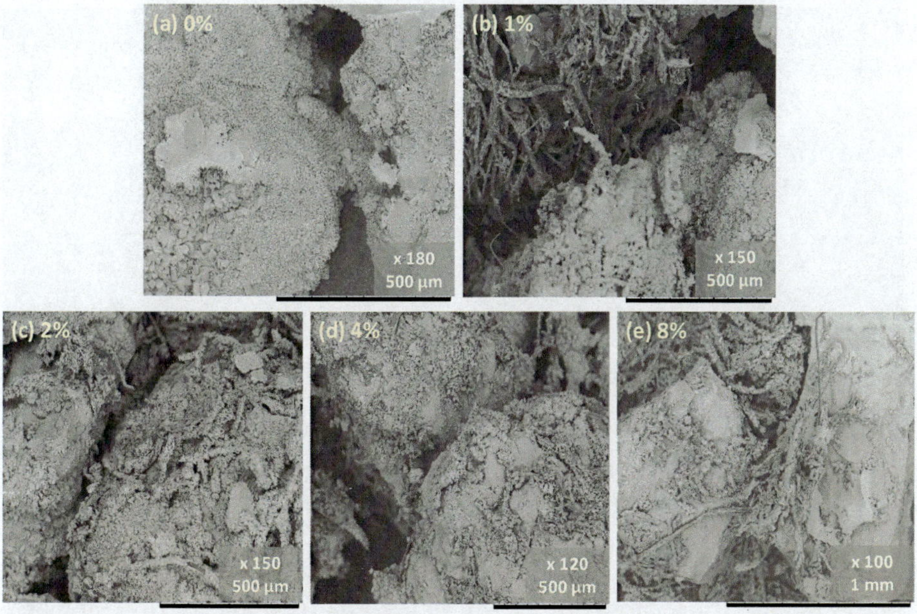

Fig. 7. SEM images of MICP treated samples

However, a small quantity of fiber addition increases carbonate production as it acts as a filter to maintain the bacteria population in samples. The optimum fiber addition, which balances the strength gain and the failure strain, was found to be 1% of fiber addition. A similar tendency related to fiber addition was also observed by Li et al. (2016). This study made fiber reinforced, bio-cemented sand samples using uniform polypropylene multifilament fiber (12 mm long and 0.1 mm thick) at fiber ratios of 0.1, 0.2, 0.3, and 0.4% (by weight). Moreover, the results showed that the failure strain of MICP treated sand at the optimum fiber addition (found to be 0.2–0.3%) was 3 times higher than sand-only samples.

The 2% fiber addition case seems to be not in line with other samples. The Most likely cause of its poor performance is possibly the inhomogeneous distribution of fibers and carbonate precipitation, which leads to a blockage in the sample and a decrease in the strength gain.

3.3 Effect of Fiber on Survivability of the Bacteria

It was hypothesized that the used paper fiber enhance the bacteria growth, and the composition of natural fibers could serve as the energy sources for microorganisms. Figure 8 shows the results of bacterial survivability tests obtained from fluorescence microscopy analysis. Figure 8(a–b) presents two control cases: one for examining the bacteria on sand materials; another one for used paper fibers. It is somewhat surprising that fibers were fluorescent, which is out of previous expectations. In Fig. 8(c–d), samples after two-week treatment were illustrated. The results interestingly show that 2% fiber case fluoresces much more than 0% fiber case, which demonstrates the higher survival

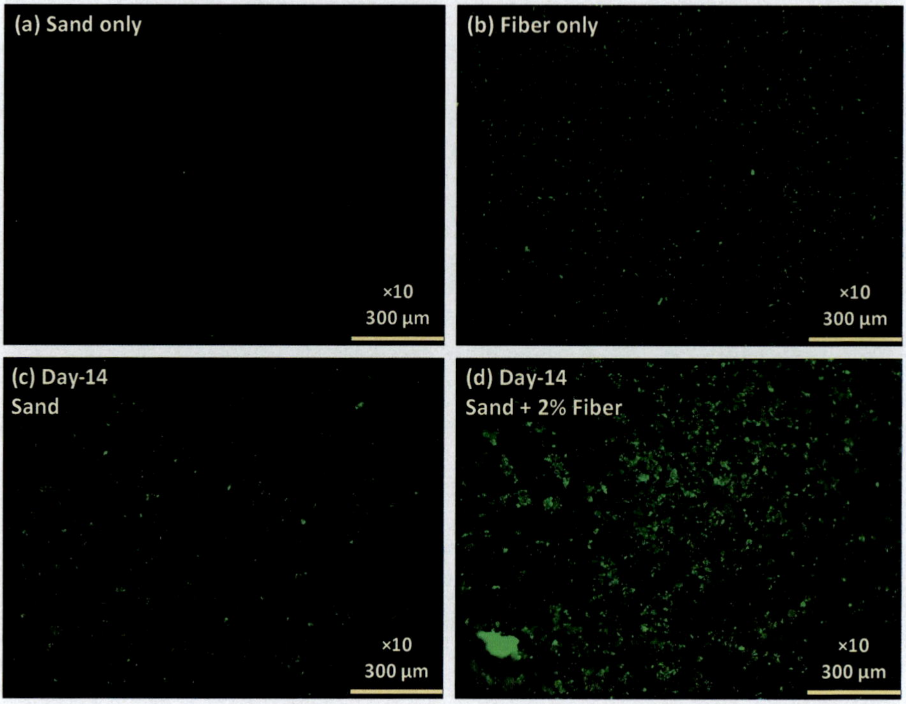

Fig. 8. Fluorescence microscopy images

of the bacteria at fiber addition. Although control cases fluoresced as well, a rough comparison could be made to state that fibers did enhance the survivability of bacteria in treated samples, and future works are highly recommended to deepen the understanding on this subject.

4 Limitations

The generalisability of these results is subject to certain limitations. For instance, the waste fiber used in this study is not uniform either in length, size, or composition, making it difficult to characterize. Another source of weakness in this study is that the change in mechanical properties was not addressed in this study; therefore, it is unknown if the used paper fiber could alter the failure pattern of traditional bio-cemented sand materials, which was confirmed as brittle materials in previous studies. In order to achieve a wider application, further experimental investigations are needed to establish a more comprehensive understanding of used paper fiber incorporated bio-cementation, such as uniaxial compression test, durability test, and so on. Notwithstanding these limitations, this work provides valuable insights into the solidification mechanism of the used paper fiber incorporated bio-cementation. Overall, facts have proved that as an environmental-friendly option, the used paper fiber has the potential to contribute to a global trend of sustainable development.

5 Conclusions

In this investigation, the aim was to evaluate the feasibility of used paper fiber incorporated MICP on sands and how the fiber material could enhance the micp process in sand. For this purpose, three sets of experiments were carried out; the results later confirmed that the used paper fiber addition could remarkably increase the yield of carbonate precipitation and improve the strength. The main significant findings were concluded below.

The used paper fiber studied in this research could serve as the energy source for bacteria, thus enhance the survivability of bacteria and increase the yield of carbonate precipitation. Furthermore, the fiber plays a role as a filter inside treated samples. When the fiber addition was up to 2%, more than 50% of bacteria could remain in samples.

The distribution of carbonate precipitation was greatly influenced by the fiber addition since the "fiber filter" made more bacterial stay at the top part of the samples. Therefore, the carbonate content of the top part was improved as 1.5 times as the rest part.

A small quantity of fiber addition was proved to be enough for the strength improvement, and the optimum was found to be 1%, which doubled the estimated UCS (compared with no fiber case) in this study. However, if the addition goes up to 2%, the strength gain would be down to less than cases without fibers.

The SEM images indicated that not only the yield of carbonate precipitation but also the effective bonding should be taken into consideration when analysing the solidification mechanism. The loss of adequate, effective bonding was considered the reason for the strength-decreasing tendency in high fiber addition cases.

References

Amarakoon, G.G.N.N., Kawasaki, S.: Factors affecting sand solidification using MICP with *Pararhodobacter* sp. Mater. Trans. **59**, 72–81 (2018). https://doi.org/10.2320/matertrans.M-M2017849

Danjo, T., Kawasaki, S.: Microbially induced sand cementation method using *Pararhodobacter* sp. strain SO1, inspired by beachrock formation mechanism. Mater. Trans. **57**, 428–437 (2016). https://doi.org/10.2320/matertrans.M-M2015842

DeJong, J.T., Mortensen, B.M., Martinez, B.C., Nelson, D.C.: Bio-mediated soil improvement. Ecol. Eng. **36**, 197–210 (2010). https://doi.org/10.1016/j.ecoleng.2008.12.029

Dejong, J.T., et al.: Biogeochemical processes and geotechnical applications: progress, opportunities and challenges. Geotechnique **63**, 287–301 (2013). https://doi.org/10.1680/geot.SIP13.P.017

Fang, X., Yang, Y., Chen, Z., Liu, H., Xiao, Y., Shen, C.: Influence of fiber content and length on engineering properties of MICP-treated coral sand. Geomicrobiol. J. **37**, 582–594 (2020). https://doi.org/10.1080/01490451.2020.1743392

Feng, K., Montoya, B.M.: Influence of confinement and cementation level on the behavior of microbial-induced calcite precipitated sands under monotonic drained loading. J. Geotech. Geoenviron. Eng. **142**, 04015057 (2016). https://doi.org/10.1061/(ASCE)GT.1943-5606.0001379

Fukue, M., Nakamura, T., Kato, Y.: Cementation on soils due to calcium carbonate. Soils Found. **39**, 55–64 (1999). https://doi.org/10.3208/sandf.39.6_55

Gowthaman, S., Iki, T., Nakashima, K., Ebina, K., Kawasaki, S.: Feasibility study for slope soil stabilization by microbial induced carbonate precipitation (MICP) using indigenous bacteria isolated from cold subarctic region. SN Appl. Sci. **1**(11), 1–16 (2019). https://doi.org/10.1007/s42452-019-1508-y

Gowthaman, S., Mitsuyama, S., Nakashima, K., Komatsu, M., Kawasaki, S.: Biogeotechnical approach for slope soil stabilization using locally isolated bacteria and inexpensive low-grade chemicals: a feasibility study on Hokkaido expressway soil, Japan. Soils Found. **59**, 484–499 (2019). https://doi.org/10.1016/j.sandf.2018.12.010

Gowthaman, S., Nakashima, K., Kawasaki, S.: Freeze-thaw durability and shear responses of cemented slope soil treated by microbial induced carbonate precipitation. Soils Found (2020). https://doi.org/10.1016/j.sandf.2020.05.012

Gowthaman, S., Nakashima, K., Kawasaki, S.: A state-of-the-art review on soil reinforcement technology using natural plant fiber materials: past findings, present trends and future directions. Materials (Basel) **11**, 553 (2018). https://doi.org/10.3390/ma11040553

Imran, M.A., Gowthaman, S., Nakashima, K., Kawasaki, S.: The influence of the addition of plant-based natural fibers (jute) on biocemented sand using MICP method. Materials **13**, 4198 (2020). https://doi.org/10.3390/ma13184198

Ivanov, V., Chu, J.: Applications of microorganisms to geotechnical engineering for bioclogging and biocementation of soil in situ. Rev. Environ. Sci. Biotechnol. **7**, 139–153 (2008). https://doi.org/10.1007/s11157-007-9126-3

Khattak, M.J., Alrashidi, M.: Durability and mechanistic characteristics of fiber reinforced soil–cement mixtures. Int. J. Pavement Eng. **7**, 53–62 (2006). https://doi.org/10.1080/10298430500489207

Li, M., Li, L., Ogbonnaya, U., Wen, K., Tian, A., Amini, F.: Influence of fiber addition on mechanical properties of MICP-treated sand. J. Mater. Civ. Eng. **28**, 1 (2015). https://doi.org/10.1061/(ASCE)MT.1943-5533.0001442

Liu, S., et al.: Enhancement of MICP-treated sandy soils against environmental deterioration. J. Mater. Civ. Eng. **31**, 1–13 (2019). https://doi.org/10.1061/(ASCE)MT.1943-5533.0002959

Naveed, M., Duan, J., Uddin, S., Suleman, M., Hui, Y., Li, H.: Application of microbially induced calcium carbonate precipitation with urea hydrolysis to improve the mechanical properties of soil. Ecol. Eng. **153**, 105885 (2020). https://doi.org/10.1016/j.ecoleng.2020.105885

Park, S.-S.: Unconfined compressive strength and ductility of fiber-reinforced cemented sand. Constr. Build. Mater. **25**, 1134–1138 (2011). https://doi.org/10.1016/j.conbuildmat.2010.07.017

Teng, F., Ouedraogo, C., Sie, Y.C.: Strength improvement of a silty clay with microbiologically induced process and coir fiber. J. Geoengin. **15**, 79–88 (2020). https://doi.org/10.6310/jog.202006_15(2).2

van Paassen, L.A., Ghose, R., van der Linden, T.J.M., van der Star, W.R.L., van Loosdrecht, M.C.M.: Quantifying biomediated ground improvement by ureolysis: large-scale biogrout experiment. J. Geotechn. Geoenviron. Eng. **136**(12), 1721–1728 (2010). https://doi.org/10.1061/(ASCE)GT.1943-5606.0000382

Zhao, Y., Fan, C., Ge, F., Cheng, X., Liu, P.: Enhancing strength of MICP-treated sand with scrap of activated carbon-fiber felt. J. Mater. Civ. Eng. **32**, 1–8 (2020). https://doi.org/10.1061/(ASCE)MT.1943-5533.0003136

A Field Study on the Utilization of Recycled Concrete Aggregate in Drainage Systems

Jinwoo An[1](✉), Adam Lane Perez[2], Boo Hyun Nam[2], and Byoung Hooi Cho[2]

[1] School of Engineering, Department of Civil and Mechanical Engineering, University of Mount Union, 1972 Clark Ave, Alliance, OH 44601, USA
anji@mountunion.edu

[2] Department of Civil, Environmental and Construction Engineering, University of Central Florida, 4000 Central Florida Blvd. Building 91, Suite 211, Orlando, FL, USA
alperez87@knights.ucf.edu, {boohyun.nam,byoungcho}@ucf.edu

Abstract. A field study of the exfiltration behavior of French drain system built with recycled concrete aggregate (RCA) as a pipe backfill material was conducted. This study was designed to evaluate the long-term exfiltration performance of four types of French drain systems in the field test. Each French drain system was constructed with different aggregate type and condition, which involve limestone, RCA 'as is', RCA with 2% and 4% additional fines. To evaluate the exfiltration performance of RCA French drain systems, field and laboratory tests were employed. From the field test, the flow rate and discharge rate of French drain systems were monitored and measured for 12 months to estimate an average permittivity of each trench in the field. Two laboratory tests of this study included exfiltration test of nonwoven geotextiles and material characterization of RCA fines were conducted. Geotextiles were excavated and taken from the four RCA French drain systems after 12 months and tested to measure the average permittivity. The results of field and laboratory tests show that major factors affecting the exfiltration performance of French drain system are permeability of surrounding soil condition and groundwater table in the field. The aggregate type (RCA or limestone), on the other hand, is a minor factor affecting the exfiltration performance of French drain system. In conclusion, the utilization of RCA as a pipe backfill material in a drainage system causes no significant reduction in exfiltration performance of a drainage system in comparison with the natural aggregate (limestone).

Keywords: Sustainability · Recycled concrete aggregate · French drain · Field study

1 Introduction

Recycled concrete aggregate (RCA) is often used as a replacement of virgin aggregate in road foundations (base course), embankments, hot-mix asphalt, and Portland cement concrete. However, the use of RCA in exfiltration drainage systems, such as French drains, is currently prohibited in many states of the U.S. Because the reuse of RCA as drainage material, for instance in base and subbase layer, has shown the potential of

S. S. Kim et al. (Eds.): GeoChina 2021, SUCI, pp. 65–89, 2021.
https://doi.org/10.1007/978-3-030-80152-6_6

clogging of filter-fabric. Generally, the presence of fine particles (or "fines" smaller than 75 μm) in the aggregate causes the clogging of filter fabric, resulting in rendering the filter inoperable, or significantly reducing its performance efficiency. The use of RCA in pavement base layers has been associated with reductions in filter fabric permittivity and the accumulation of precipitate and other materials in pavement subdrainage systems [1]. Primary performance concern is the deposition of calcite precipitate and other fines in the drainage system [2]. Some laboratory studies have shown that RCAs can produce significant calcium carbonate (calcite) deposits while virgin aggregates such as gravel and crushed stone do not [3, 4]. In addition, the fines from RCA may pose another risk if they rehydrate in the pores of the filter fabric. Recementaion of RCAs in stockpiles has been reported many times [1, 5].

A French Drain collects water from the roadway and transfers the water into slotted pipes underground. The water then filters through coarse aggregate and passes out through a permeable filter fabric. Figure 1a shows a schematic diagram of French Drain having No. 4 RCA as subdrainage aggregate. Despite those clogging potential mentioned above, the reuse of RCA in French Drains has recently received a great attention because of the requirement of large aggregate in the drain. The required aggregate is No. 4 gradation (see Fig. 1b) and it can minimize the clogging potential. Although extensive studies on the RCA used as base/subbase material and filler materials in Portland cement concrete (PCC) and hot-mix asphalt (HMA) have been conducted, the RCA in French Drain applications has not been studied in the US.

Fig. 1. Drainage system with RCA: (a) schematic diagram of French Drain and (b) FDOT No. 4 aggregate gradation

The performance of RCA as drainage material has not been widely evaluated, and the limited information limits its use. Some state highway agencies have reported the use of RCA as base course; however, no state reports the use of RCA in exfiltration drainage systems. In several previous studies of RCA as drainage material, a series of laboratory-scale permeability and clogging tests were conducted to evaluate the drainage performance of RCA No. 4 aggregate [5, 6]. In permeability and clogging tests, the changes of flow rate by the addition of RCA fine (defined as particles smaller than 75 μm) were measured over time. Testing results could provide insights of RCA material characteristics associated with short- and long-term drainage performance. In particular, the influence of RCA fines was clearly observed by reducing the outflow rate over time [5, 6]. These small-scale tests, however, have revealed some limitations summarized as below:

- The permeability testing simulates one-dimensional flow only.
- A scale effect cannot be avoided because a small permeameter (9-in. diameter) and a large size of aggregates (No. 4 aggregate) were used.
- Calcite precipitation in the clogging test cannot represent in-situ condition.

Therefore, it is necessary to setup in-situ/full-scale experiments that simulate three-dimensional flow and to make long term monitoring of performance. The measurements and conclusions will be used to develop or modify Specifications for the Use of RCA in French Drains.

2 Objectives

In this study, RCA was tested for its physical properties, including specific gravity, unit weight, percent voids, absorption, and abrasion resistance. And, constructed in-situ and full-scale French drain systems are introduced with different parameters, such as, percent fines addition (0, 2 and 4% fines of RCA, % of fines are by weight) and drainage condition. The objectives of this research on the in-situ study of RCA French drain are:

1) to develop field test methods to measure the drainage performance of in-situ and full-scale French Drain, and
2) to monitor long-term flow rate of French drain to evaluate the tendency of geotextile clogging due to RCA fines and
3) to investigate long-term drainage performance and clogging buildup in the in-situ French Drain systems by the discharge rate of groundwater.

3 Materials

3.1 Materials

Excluding the perforated drainage pipe, a French drain system has three materials that restrict the flow of water, they are the aggregate (limestone or RCA in this experiment), geotextile, and in-situ soil. The performance of the drainage system from a materials perspective is most sensitive to the properties of the in-situ soil compared to the aggregate or geotextile since the in-situ soil is the most restrictive element, thus controlling the equivalent hydraulic conductivity of the French drain system.

3.2 Materials Properties of Limestone

Limestone (used in the control trench) has been tested and its material properties are presented herein. Figure 2 shows the photo of limestone as received. Sieve analysis (AASHTO T 27/T 11/M 145) was conducted and the results are presented in Table 1 and Fig. 3.

In addition, L.A. abrasion (ASTM C535) and specific gravity (AASHTO T 85) tests were conducted on the limestone. The L.A. abrasion of 28.9% was measured and different types of specific gravity which were determined at 68 °F are presented in Table 2. The bulk density and voids in limestone aggregate were measured according to AASHTO T 19 and the results are also shown in Table 2.

Fig. 2. Limestone aggregate

Table 1. Sieve analysis of fine and coarse aggregates (AASHTO T 27/T 11/M 145)

Percent finer (%)	Diameter (in.)	Classification	Cu	Cc
D60	1.190	A-1-a	1.51	1.04
D30	0.988			
D10	0.790			

Note: (conversion: 1 in. = 25.4 mm), Where, D60 = diameter corresponding to 60% finer; D30 = diameter corresponding to 30% finer; D10 = diameter corresponding to 10% finer; Cu = uniformity coefficient; Cc = coefficient of gradation

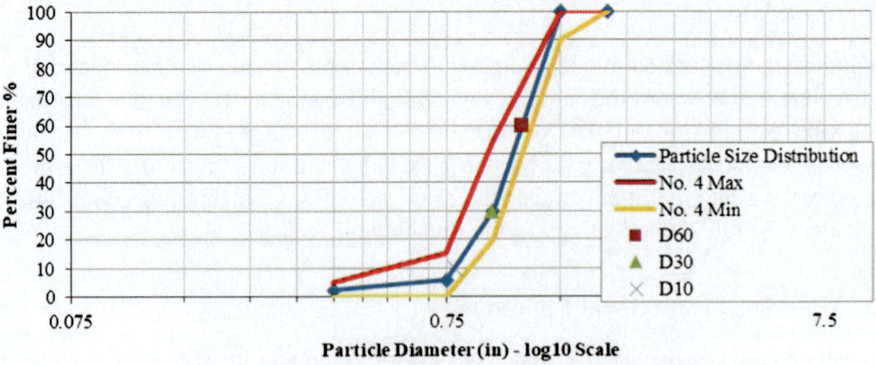

Fig. 3. Particle size distribution of limestone aggregate

Table 2. Specific gravity, absorption, bulk density and void content of limestone aggregate (AASHTO T 85)

Bulk Dry SG.	Bulk SSD SG.	Apparent SG.	Absorption (%)	Bulk density (lb/ft^3)	Void content (%)
2.31	2.39	2.51	3.5	80.5	44.1

Note. conversion: 1 lb/ft^3 = 16.0185 kg/m^3

3.3 Materials Properties of RCA

The RCA has been tested and its material properties are presented below. Figure 4 shows the photo of RCA as received. Sieve analysis (AASHTO T 27/T 11/M 145) was conducted and the results are presented in Table 3 and Fig. 5.

Fig. 4. RCA aggregate

Table 3. Sieve analysis of RCA aggregates (AASHTO T 27/T 11/M 145)

Percent finer (%)	Diameter (in.)	Classification	Cu	Cc
D60	1.190	A-1-a	1.49	0.988
D30	0.988			
D10	0.790			

In addition, L.A. abrasion (ASTM C535) and specific gravity (AASHTO T 85) tests were conducted on the RCA aggregate. The L.A. abrasion of 43.7% was measured and specific gravities are presented in Table 4. The bulk density and voids in RCA aggregate were measured according to AASHTO T 19 and the results are also shown in Table 4.

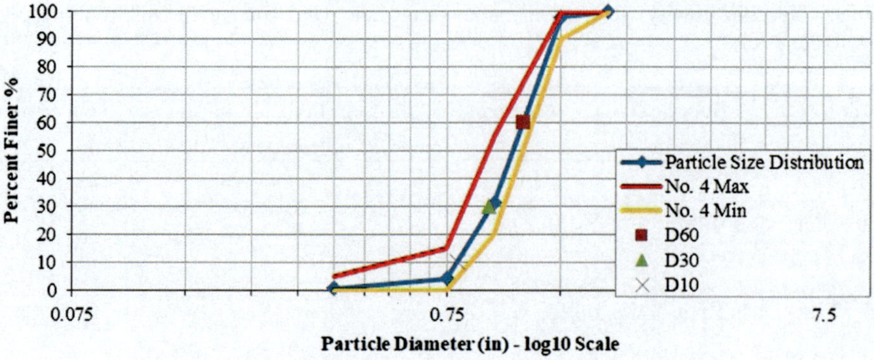

Fig. 5. Particle size distribution of RCA aggregate

Table 4. Specific gravity, absorption, bulk density and void content of RCA aggregate (AASHTO T 85)

Bulk dry SG.	Bulk SSD SG.	Apparent SG.	Absorption (%)	Bulk density (lb/ft^3)	Void content (%)
2.02	2.21	2.50	9.6	70.8	43.7

3.4 Properties of In-Situ Soils

Materials properties of the in-situ soils were tested. The testing methods include the sieve analysis (AASHTO T 27/T 11/M 145), specific gravity (D854 Method B), and permeability (ASTM D2434). The in-situ soils are composed of three layers as shown in Fig. 6. Table 5 shows the results of sieve analysis for soils in Layers 1, 2, and 3. The specific gravity and permeability of those soils are presented in Table 6.

Fig. 6. Photo showing the three layers of soil at the site

Table 5. Sieve analysis of soils for layers 1, 2, and 3 (AASHTO T 27/T 11/M 145)

Layer	Percent finer (%)	Diameter (mm)	Classification	Cu	Cc
Layer1	D60	0.200	A-3	1.89	1.05
	D30	0.149			
	D10	0.106			
Layer2	D60	0.198	A-3	1.85	0.99
	D30	0.145			
	D10	0.107			
Layer3	D60	0.197	A-3	1.82	1.00
	D30	0.145			
	D10	0.108			

Table 6. Permeability of soils from Layers 1, 2, and 3 (ASTM D2434)

Layer	Constant head @ 68 °F (ft/s)	Falling head @ 68 °F (ft/s)	Average @ 68 °F (ft/s)	Moist unit weight (lb/ft^3)	Dry unit weight (lb/ft^3)	Void ratio
Layer 1	$1.48*10^{-4}$	$1.23*10^{-4}$	1.35*10–4	92.2	84.5	0.95
Layer 2	$1.49*10^{-4}$	$1.33*10^{-4}$	$1.41*10^{-4}$	105.1	93.4	0.74
Layer 3	$1.96*10^{-5}$	$2.16*10^{-5}$	$2.06*10^{-5}$	107.6	89.0	0.84

3.5 Material Properties of Geotextile

Geotextile (N060 polypropylene nonwoven fabric) was purchased and its properties measured by the manufacturer are summarized in Table 7. The geotextile permittivity prior to installation has a minimum average roll value of 1.6 s^{-1}. Drainage geotextiles as required by Section 985 Specifications for FDOT Road and Bridge Construction must have a minimum permittivity of 0.5 s^{-1} (D-3a) for the least restrict D-3 geotextile. Testing of the geotextile was conducted by the FDOT State Materials Office and it was concluded that the samples of the geotextile provided meet all the requirements for type D-3 French drain filter fabric as specified in 2016 FDOT Standard Specifications for Road and Bridge Construction, Section 985.

Table 7. Technical data sheet N060 nonwoven geotextile

Property	Minimum average roll value	Test method
Grab tensile strength	160 lbs	ASTM D4632
Grab tensile elongation	50%	ASTM D4632
CBR Puncture	410 lbs	ASTM D6241
Trapezoid tear strength	60 lbs	ASTM D4533
UV Resistant @ 500 h	70%	ASTM D4355
Apparent opening size (AOS)	70 US Sieve	ASTM D4751
Permittivity (sec-^{1})	1.6 (s^{-1})	ASTM D4491
Flow rate	110 gpm/ft2	ASTM D4491

3.6 French Drain and Monitoring System Design

The design of the in-situ French drain system is shown in Fig. 7. A total of four French drain systems and twenty-five monitoring wells were constructed at the project site. Each French drain contains a different aggregate composition, from east to west the four French drains are composed of RCA "as is", RCA with 2% fines, RCA with 4% fines, and virgin limestone as the control. For instance, Trench 1 consists of RCA 'as is'. Trench 2 and 3 are made of RCA with the additions of 2% and 4% fines of RCA, respectively. Trench 4 which is the control is composed of limestone. The French drains were constructed in the east to west order, with the monitoring wells constructed subsequently.

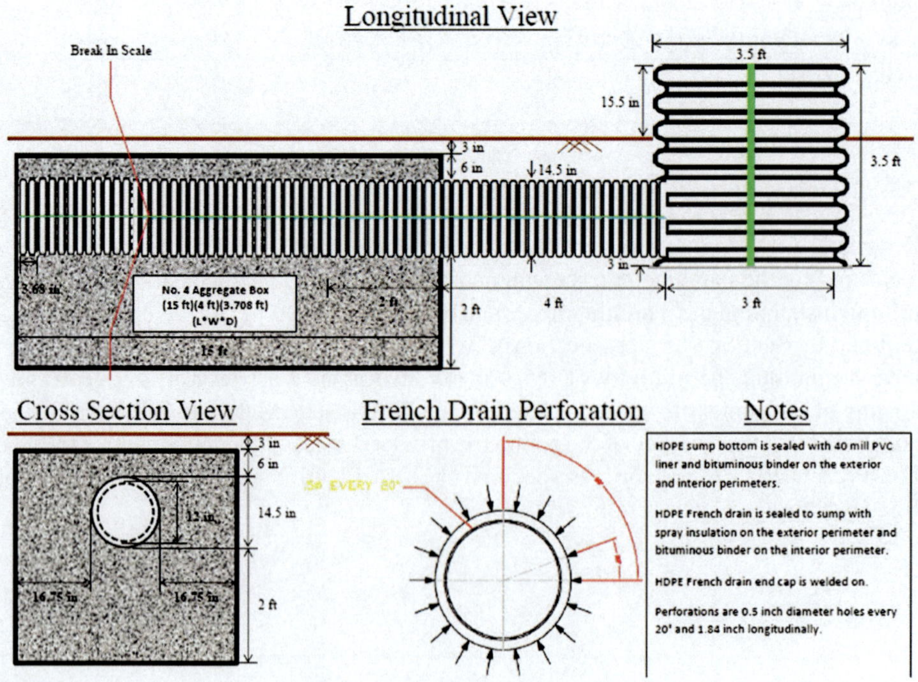

Fig. 7. Design of the in-situ French drain system

All four French drain trenches were excavated using shovels with the spoil being placed adjacent to the trenches (see Fig. 8). Three distinct volumes were excavated within a French drain trench, 1) the volume removed for the drainage aggregate, approximately 15 ft * 4 ft * 4 ft, 2) the volume removed for the pipe segment between the sump and drainage aggregate, approximately 4 ft * 2 ft * 2 ft, and 3) the volume removed for the sump, approximately 5 ft * 5 ft * 2 ft. The different volumes of excavation are shown in Fig. 8. As RCA French drain trench was being excavated it also served as a test trench to collect in-situ soil samples. Three soil layers were identified along the depth of the trench and soil samples were obtained by pushing the acrylic tube portion of permeameter cells

Fig. 8. Underground trench excavation: (a) Trench 1 (RCA 'as is'), (b) Trench 2 (RCA with 2% fines), (c) Trench 3 (RCA with 4% fines), and (d) Trench 4 (limestone)

into the side of the trench by means of a mechanical jack. The acrylic tubes were then taken to the lab and reinstalled into the respective permeameters. Laboratory testing of the in-situ soil samples is outlined in the previous section (see *properties of in-situ soils).*

3.6.1 Water Delivery System

The fire hydrant on site is rated orange (500–999 GPM) and serves as the water supply when running the French drain experiment. The flowmeter was installed (see Fig. 9) to measure the flow rate during experimental test. The fire hose connects to the flowmeter and supplies water to the individual French drains.

(a) Fire hydrant with flowmeter and backflow preventer **(b) Flowmeter**

Fig. 9. On site water delivery system

3.6.2 In-situ Water Level System and Sensors

The in-situ groundwater monitoring system for an individual French drain consists of seven HDPE monitoring wells which collect data on the groundwater mounding profile. Each well contains a data logging piezometer suspended by a wire connected to the well cap. These monitoring wells are placed along the lateral and axial centerline of the aggregate box. The wells are constructed out of 4 in. inner diameter perforated pipe and placed into the ground 7 ft deep. Figure 10 shows the piezometer sensor and the schematic diagram of the piezometer sensor installation in a monitoring well. The wells placed at 7 ft laterally between French drains marks the halfway point between drains. In addition to variation in flow rates, change in the groundwater mounding profiles are used to quantify the performance of the French drains.

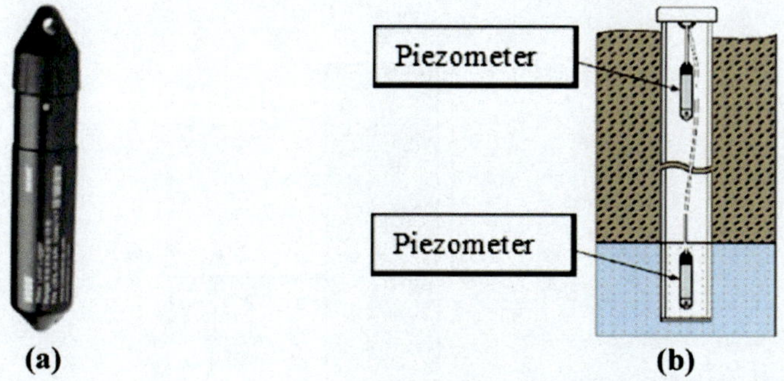

Fig. 10. Water level sensor: (a) piezometer sensor (HOBO U20L-04) and (b) sensor installation

The site plan view showing the locations of four French drains and the fire hydrant which is the source of water supply is presented in Fig. 11. The schematic diagrams showing the water level measuring system design are presented in Figure 12(a) and (b) (top and cross-sectional views, respectively). Three monitoring wells are at a distance of 1 ft from the edge of the French drain to the centerline of the well, with one being at the tip of the drain and the other two being along the lateral center line of the drain, with one well on each side of the drain. Four additional wells are placed along the lateral center line of the drain at 3 ft increasing increments away from the 1 ft well, with two of these additional wells being on each side of the drain. If the outermost lateral well (7 ft from the French drain) is the half-way point between two drains, then this well will be shared between the well systems of both drains. The data logging piezometers are HOBO U20L-04 model, a single data extracting device known as the HOBO Waterproof Shuttle model (U-DTW-1) was also purchased, as well as the data analysis software labeled HOBOware PRO v.3.x for PC & Mac item number (BHW-PRO-CD).

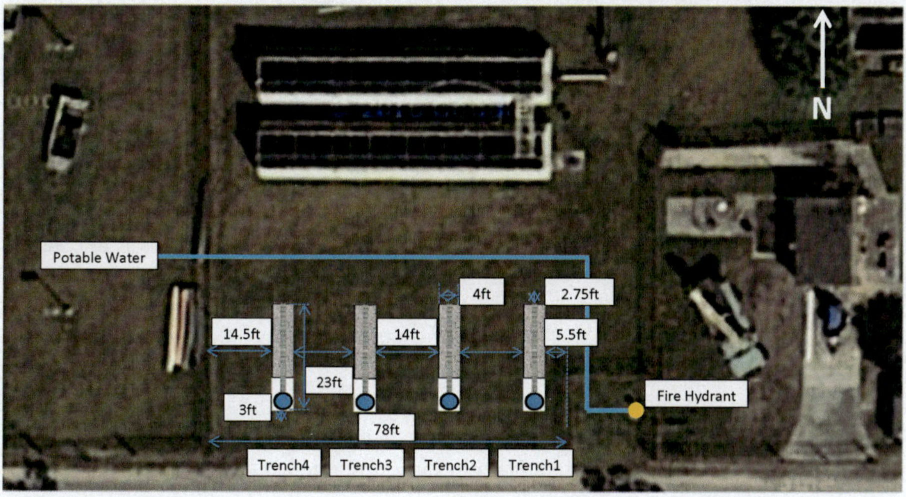

Fig. 11. French drain site plan view

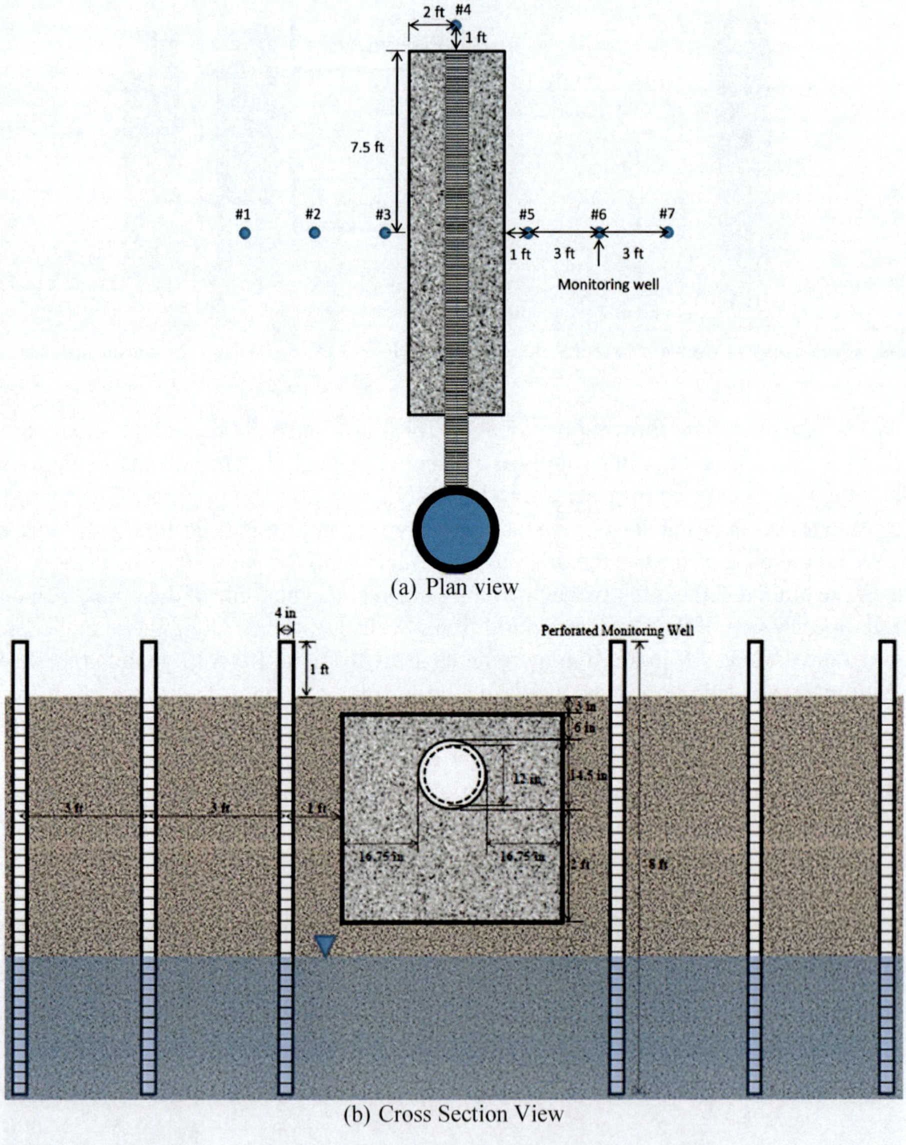

Fig. 12. Design of in-situ water level measuring system: (a) Field water level system plan view and (b) cross-sectional view of French drain with dimensional information

3.6.3 Experimental Procedure of In-situ RCA French Drain

1) With the water level meter, calculate the depth of the initial groundwater table in each monitoring well relative to the ground surface before testing.
2) Launch the piezometers with a 30-min sampling rate for the French drain which is about to be tested.
3) Place the piezometers in the monitoring wells of the French drain under testing and ensure the piezometers are fully submerged under the initial groundwater table.

4) Allow the piezometers to take one measurement before testing so the positions of the piezometers relative to the initial groundwater table can be back calculated through the increase in absolute pressure above atmospheric pressure (14.7 psi).
5) Install the hose to the flowmeter and place the other end into the sump.
6) Turn on the water.
7) Obtain a head of water in the sump equal to the elevation of the ground surface (head difference across the French drain system is then equal to the difference between the initial groundwater table and ground surface) within the first 10 min and maintain a constant head at that level for 80 min
8) After the 80 min (90 min in total) collect flow rate data from the flowmeter.
9) Turn off the water supply
10) Allow the piezometers to measure the behavior of the groundwater mounding dissipation over a minimum of 72 h, collect data later on before testing of the next French drain
11) Put away hose and water level meter

4 Results and Discussions

4.1 French Drain Performance

4.1.1 Flow Rate of French Drain

The flow rate measurements were taken after 1.5 h of supplying water to the French drain system. Flow rate values for each test are shown in Fig. 13. The data has high variability which may be attributed to several factors such as soil moisture conditions before testing, direction of groundwater and time in between tests. Moisture conditions before testing can be affected by any preceding rainfall events and moisture changes in the soil profile resulting from running neighboring French drain system days prior. The red box around the data after May 1, 2018 indicates the start of the "wet" season. When performing data analysis, it was noticed that rainfall (a near daily occurrence) during the wet season would highly effect soil moisture conditions and ground water recharge rates, which in return would affect the flow rates and groundwater mounding characteristics of the French drain systems. In most of the data analysis the "wet" season data was excluded since the rainfall would cause a signification deviation from the behavior of the remaining data.

It can be seen by the data in Fig. 13 that generally the highest flow rate drains within the groundwater table range tested, are RCA 2% fines, RCA 'as is' (0.8%), RCA 4% fines, and limestone (2.2% fines), in decreasing order. At first this seems to be a contradiction of the idea that more fines would cause decreased performance. A plausible explanation for this is that RCA 2% fines and RCA 4% fines are interior drains, whereas RCA 'as is' and limestone are exterior drains. Interior drains have neighboring drains on both sides, whereas exterior drains have only a neighboring drain on one side (see Fig. 11). Interior drains thus have assisted double-sided drainage, whereas exterior drains only have assisted drainage to one side. When looking at the data with these drainage conditions in mind the data then makes sense. Figure 14 shows the flow rate

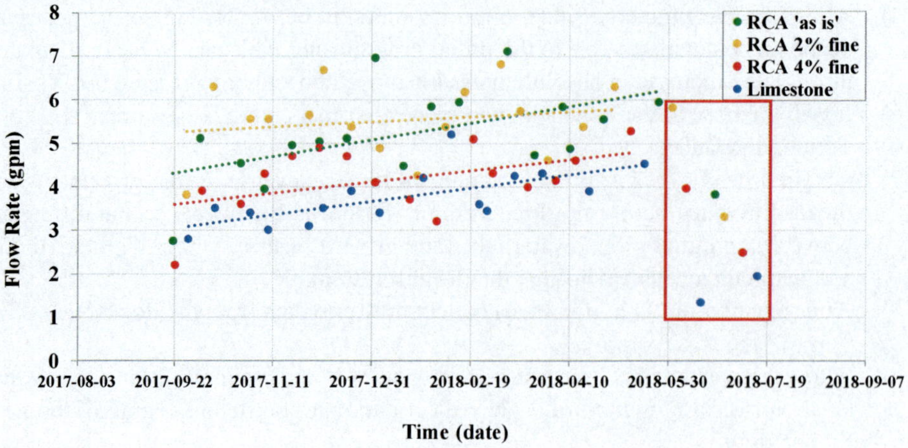

Fig. 13. Flow rate measurements with time

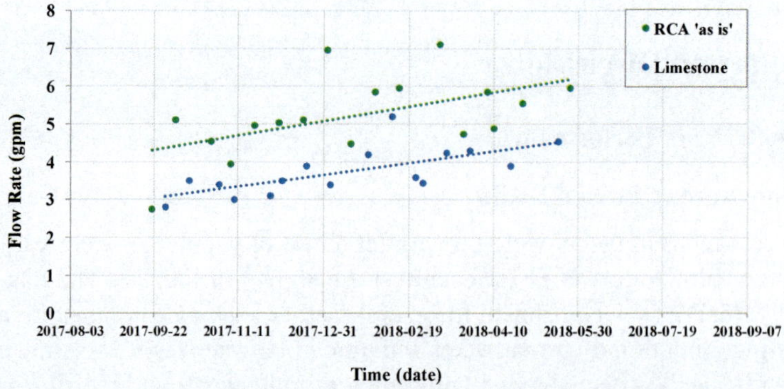

(a) Flow rate measurements of interior drains

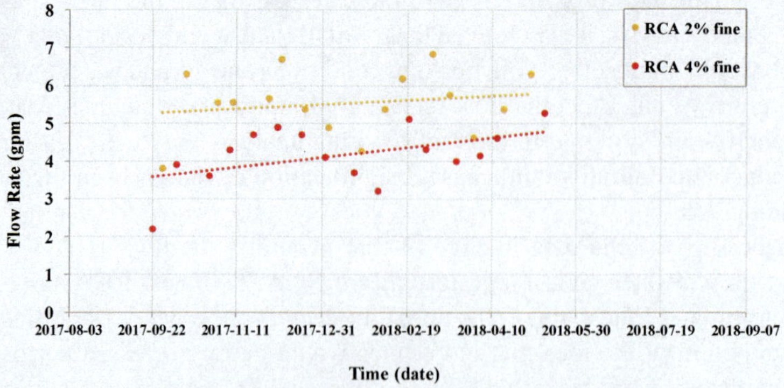

(b) Flow rate measurements of exterior drains

Fig. 14. Flow rate measurements with different drainage conditions: (a) interior drain and (b) exterior drain values of (a) interior drains and (b) exterior drains. Even though there are more fines present in RCA 2% fines than RCA 'as is', the double-sided drainage give the RCA 2% fines French drain an advantage. When comparing RCA 2% fines with RCA 4% fines (another interior French drain) it is clear that more fines do cause a reduction in flow rate. When comparing the exterior French drains it can be seen that RCA 'as is' with 0.8% fines has a higher flow rate than limestone with 2.2% fines.

The flow rate versus time trendlines for the French drains have a positive slope over the period of experimental testing. This gradual increase in flow rate is due to the seasonal change in the groundwater table depth, Fig. 15. Groundwater table depth was measured from the ground surface. As the groundwater table drops further beneath the bottom of the French drains it causes capillary forces on the unsaturated soils to increase, also the head difference between the groundwater table and water surface in the French drain increases, both mechanisms result in an increase in flow. Once again it can be seen by the data that the wet season starts around May 1st.

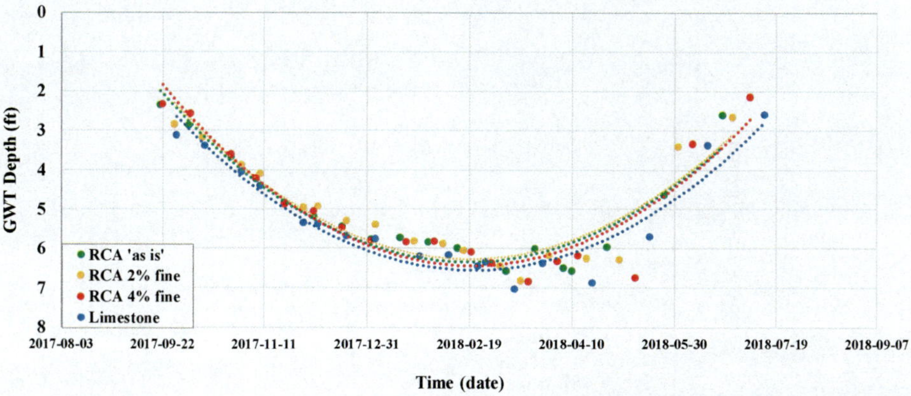

Fig. 15. Groundwater table (GWT) depth with time

Flow rate versus groundwater table depth of the French drains are plotted in Figs. 16(a) and 16(b) with the exterior and interior drains separated. Again, this separation allows for French drains with similar drainage conditions to be compared, Figs. 16(a) and 16(b), allowing variation in flow rate behavior to be isolated to differences in fine content. Both exterior French drains seem to have a similar response to changes in the groundwater table depth, as indicated by the near parallel slopes of the data. The interior French drains also share similar slopes with each other, but the slopes are flatter when compared to the exterior drains. Within the range of groundwater table depth tested, the interior drains seem less sensitive to changes in head difference. Regardless of a French drain being exterior or interior, it can be seen by both comparisons that more fines cause a reduction in flow rate. A difference in flow rate can be seen by the size of the vertical offset between the trends in both data groups. The slope of the data trends seems to be a factor of the French drain drainage conditions, whereas the vertical offset of the lines seems to be a factor of the fine content.

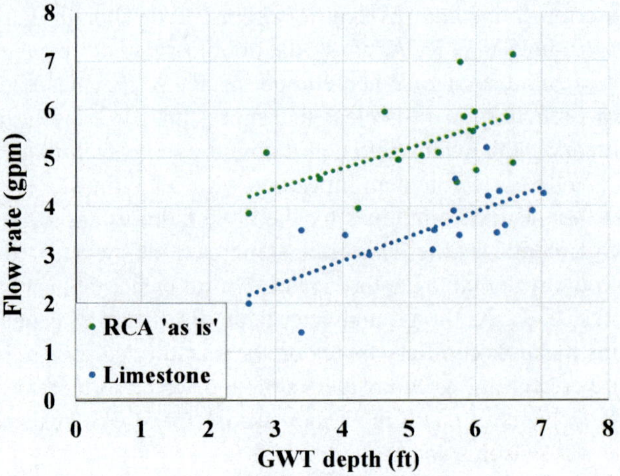

(a) Interior drain

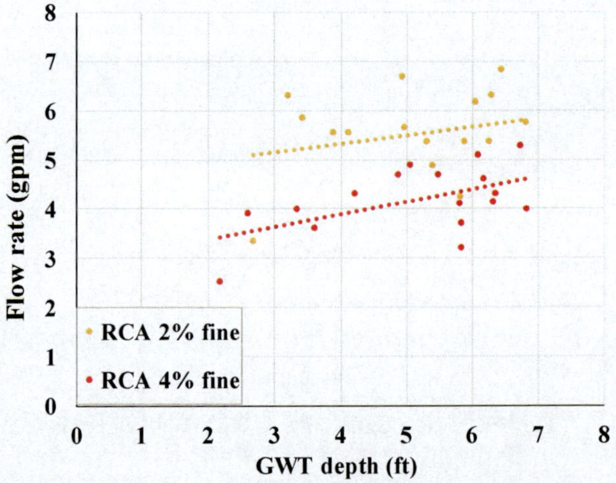

(b) Exterior drain

Fig. 16. Relationship between Flow rate and GWT depth with different drainage conditions

If the relationship between flow rate and groundwater table depth was truly linear throughout the spectrum of different groundwater table depths (i.e. head differences) then the trends should intersect the origin, since there should be no flow (i.e. 0 gpm) at a head difference of 0 ft. As can be seen by the trends in Fig. 16, only limestone even comes close to intersect the origin. In addition, there should be a maximum flow rate corresponding to a "terminal depth" of the groundwater table. The terminal depth of the groundwater table would be the depth at which an increase in depth no longer corresponds to an increase in flow, in other words, the groundwater table is so deep it no longer influences the flow rate. Mathematically, the maximum flow rate would be represented as a horizontal asymptote on the flow rate vs groundwater table depth graph. The type of function that would best represent this behavior is known as a "terminal velocity" function and is logarithmic in nature. This function allows for an origin intercept and also allows the curve to decay into a horizontal asymptote. The function is based on the following equation. $Q = Q_{avg-max}(1 - EXP(-\alpha * \Delta h))$. Where Q is the flow rate, $Q_{avg\text{-}max}$ is the average maximum flow rate (i.e. horizontal asymptote), α is a shape parameter that indicates how fast the flow rate reaches the average maximum value, and Δh is the head difference (i.e. groundwater table depth).

A curve fitting method was derived to develop a function to best represent the flow rate versus groundwater table depth relationship of each French drain. This curve fitting method allows for flow rates outside of the range of testing to be extrapolated. The first step in the curve fitting method was to apply a logarithmic trend line to the raw data. Next, the same groundwater table depth (Δh) values as the raw data where placed into the equation $Q = Q_{avg-max}(1 - EXP(-\alpha * \Delta h))$. $Q_{avg\text{-}max}$ and α values where at first set as arbitrary best guesses. After plotting the values, a logarithmic trendline was then also applied to the "curve fitting" data. Next, the parameters $Q_{avg\text{-}max}$ and α were adjusted until both logarithmic trendlines aligned on top of each other. Finally, after the values of $Q_{avg\text{-}max}$ and α were found for an individual French drain the relationship was extrapolated, Fig. 17. The parameters for each French drain are shown in Table 8. $Q_{avg\text{-}max}$ values are the average maximum values, based on initial soil moisture conditions the value has a variation in each direction of approximately 1 gpm, as can be seen by the data for each French drain. Linearization in Figs. 16(a) and 16(b) provides a general picture of the behavior of the flow rate versus groundwater table depth relationship within the testing range, whereas logarithmic extrapolation provides a more holistic view of the behavior through the spectrum of groundwater table depths.

Figure 17 shows the exterior and interior French drains pair together. Like Fig. 16, both pairs of drains exhibit similar trends. The interior drains seem to flow better at a lower groundwater table depth when compared to the exterior drains, but the interior drains are quicker to decay in flow rate and flatten out with an increasing groundwater table depth. Again, the vertical offset between the drains of similar drainage conditions can be attributed to fine content, whereas the general shape of the curve when comparing between exterior and interior drainage groups can be attributed to the drainage conditions. To compare all four French drains together regarding fine content, a correction factor would have to be developed to consider the drainage condition impact on $Q_{avg\text{-}max}$ and α. To develop such a correction factor at least two additional French drains would be needed, one exterior and one interior with the same material and fine content. Unfortunately, no such drains are present in the experimental setup, so the coupled effect of fines and drainage conditions cannot be separated from the $Q_{avg\text{-}max}$ and α. Only French drains of like drainage conditions can be compared as this allows the impact on drainage to be a constant amongst the pairs.

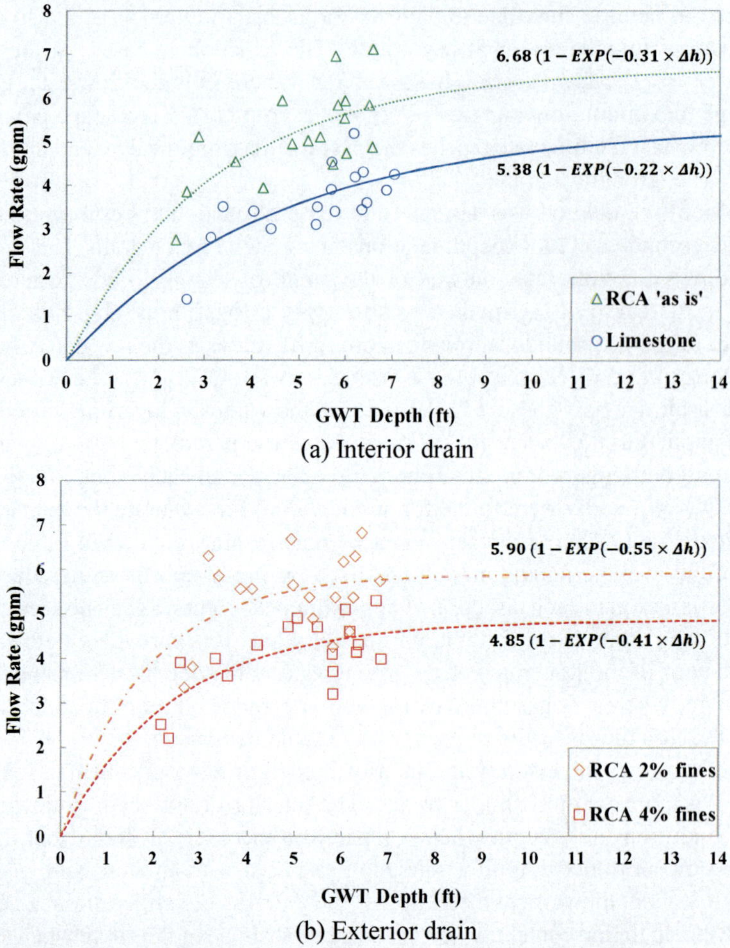

Fig. 17. Flow rate versus GWT depth relationship with different drainage conditions

Table 8. $Q_{avg\text{-}max}$ and α values of trenches

French drain	$Q_{avg\text{-}max}$ (gpm)	α
RCA 'as is'	6.68	0.31
RCA 2%	5.90	0.55
RCA 4%	4.85	0.41
Limestone	5.38	0.22

4.1.2 Discharge Behavior of French Drain (Secant Modulus)

The storage discharge behavior of each French drain was quantified by performing analysis on the monitoring wells located next to the trench sidewalls. Wells 3 and 5 were chosen because they are located 1 ft from the sides of the trench, thus the water level in these monitoring wells should closely resemble the water level within the French drain system. The rate of drop in the monitoring wells, as representing the discharge of the storage within the French drain, was calculated by measuring the head drop in both wells over a specific time interval and averaging the values together. Two time intervals were chosen for the measurement: from 2 h to 3 h and from 4 h to 5 h after the flow test began. The first interval (2 h to 3 h), designated as Stage 1, was selected as this corresponded to the exfiltration behavior 30 min after the water supply to the trench was turned off. This 30-min waiting period allowed the full storage volume to shift to a stable discharge. The second interval (4 h to 5 h), designated as Stage 2, was selected as this provided information about how the trends changed as the water level within the trench lowered. In addition, the two stages covered the measurements when the water level in the drain was at both an upper level (Stage 1) and lower level (Stage 2) relative to the French drain bottom (Fig. 18).

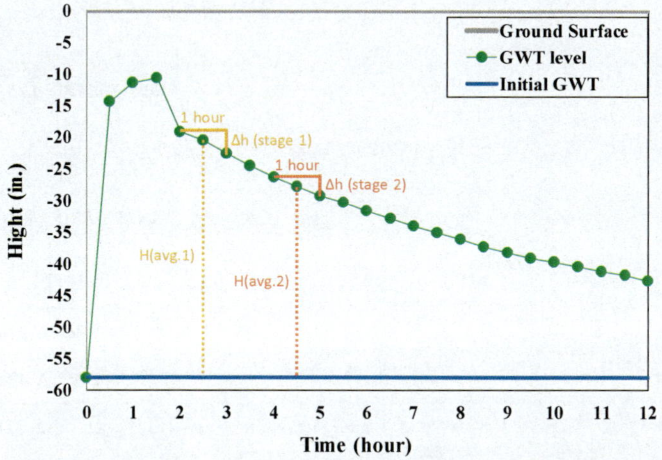

Fig. 18. Methodology for discharge behavior (secant modulus)

In Figs. 19 and 20, the secant slope is plotted against initial GWT depth. Figures 19 and 20 indicate that as time advanced throughout the experimental period the GWT depth increased, making time and initial GWT depth strongly correlated. Once again, the interior drains had better discharge performance near full storage than the exterior drains, and as the water level decreased the discharge rates of the interior drains deteriorated faster than the exterior drains. As the water level decreased (Fig. 20), the Stage 2 trends with the steepest slopes in decreasing order were, RCA 'as is', RCA with 2% fines, limestone, and RCA with 4% fines. This was also the order of increasing fines content, reinforcing that more fines caused a decrease in average drain permittivity, thus reducing the drain's slope (i.e. response to a change in head) in the relationship secant slope versus initial GWT depth. The secant slope versus initial GWT depth relationship was similar to the relationship between discharge velocity (v) versus hydraulic gradient (i), where the slope of the relationship (determined by resistance in the system) was the

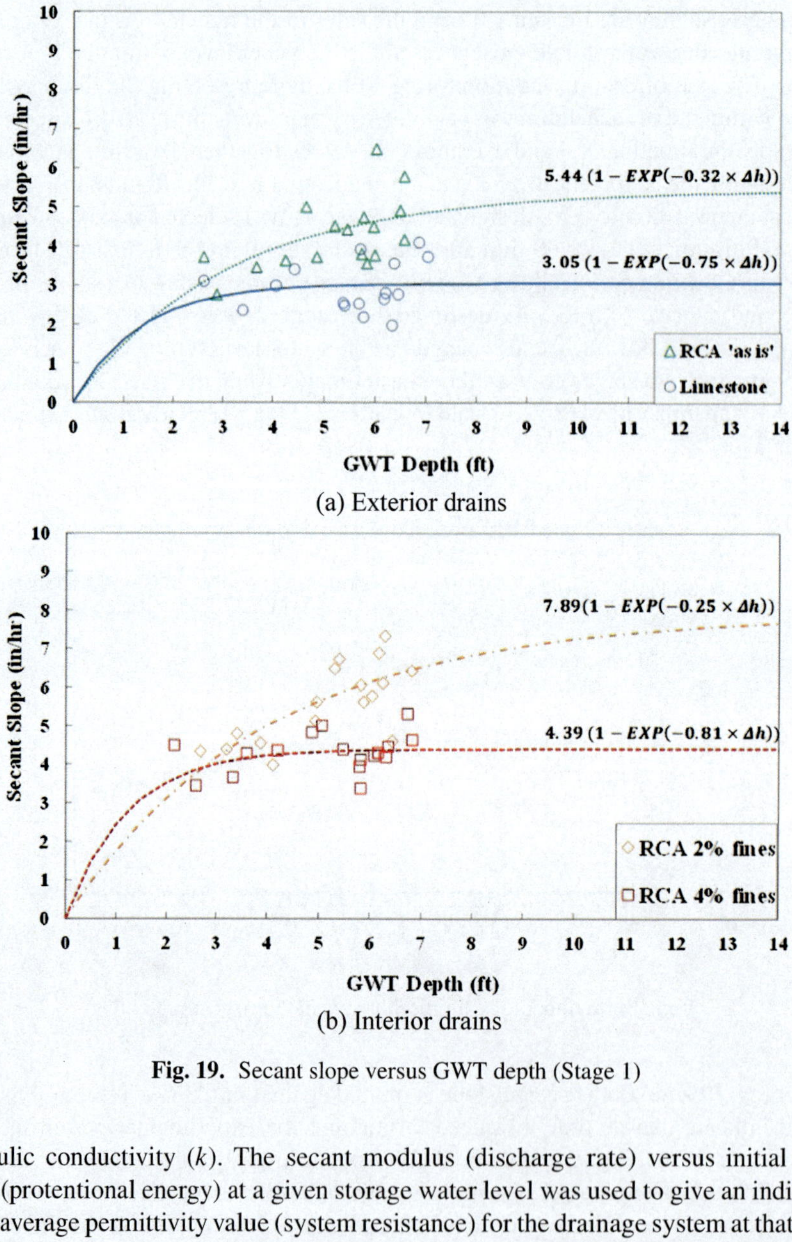

(a) Exterior drains

(b) Interior drains

Fig. 19. Secant slope versus GWT depth (Stage 1)

hydraulic conductivity (k). The secant modulus (discharge rate) versus initial GWT depth (protentional energy) at a given storage water level was used to give an indication of the average permittivity value (system resistance) for the drainage system at that water level.

The discharge rate also was estimated by the following equation below. Mathematically, the maximum discharge rate is represented as a horizontal asymptote on the plot of discharge rate versus GWT depth,

$$S = S_{avg\_max}(1 - EXP(-\alpha \times \Delta h))$$

where, S = discharge rate, S_{avg_max} = average maximum discharge rate (i.e. horizontal asymptote), α = shape parameter that indicates how fast the discharge rate reached the average maximum value, and Δh = head difference (i.e. groundwater table depth).

The parameters for each French drain are shown in Table 9. S_{avg_max} values are the average maximum discharge rate values, based on initial soil moisture conditions the value has a variation in each direction of approximately 1 in./hr, as can be seen by the data for each French drain.

Table 9. $S_{avg\text{-}max}$ and α values of French drains

Stage	French drain	$S_{avg\text{-}max}$ (in./hr)	α
1	RCA 'as is'	5.44	0.32
	Limestone	3.05	0.75
	RCA with 2% fines	7.89	0.25
	RCA with 4% fines	4.39	0.81
2	RCA 'as is'	4.81	0.32
	Limestone	2.91	0.74
	RCA with 2% fines	4.39	0.21
	RCA with 4% fines	2.46	0.31

4.2 Permittivity Testing Results of the Filter Fabric After 12 Months

Upon completion of the groundwater mounding discharge and flow rate monitoring phase of the French drain experiment, parts of each French drain were exhumed, and geotextile samples were taken. Four samples from each trench (top, middle, bottom of the side and underneath) were obtained for the permittivity testing. Geotextile samples taken from the in situ French drains are present in Fig. 21.

Comparing the permittivity values of the in situ and unused geotextile samples can provide insight into the level of clogging within each French drain. Four exhumed geotextile samples were tested for each French drain. Three flow rates per exhumed geotextile sample were used to calculate an average permittivity value for a given geotextile. Table 10 summarizes the permittivity values of each sample taken from the field.

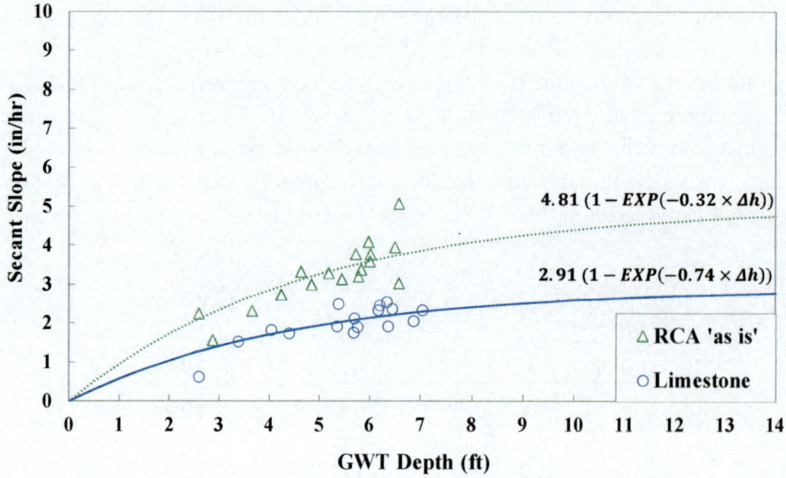

(a) Exterior drains

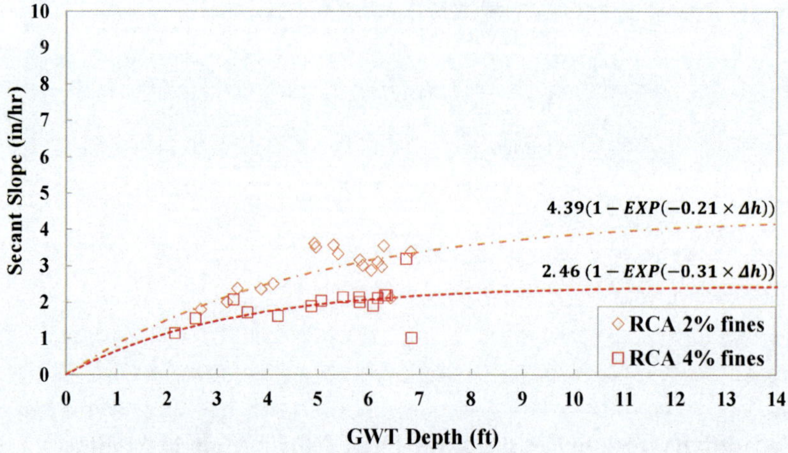

(b) Interior drains

Fig. 20. Secant slope versus GWT depth (Stage 2)

Permittivity ratio was proposed to quantify the level of "clogging" in geotextiles. Permittivity ratio was defined as the ratio of the permittivity of the exhumed geotextile to that of the permittivity of the control geotextile (1.10 s^{-1} in this study). The permittivity ratio of each exhumed geotextile can be calculated by using the equation below:

$$\text{Permittivity ratio } (\%) = \frac{Permittivity\ of\ In\text{-}situ\ Geotextile*}{Permittivity\ of\ Unused\ Geotextile}$$

(* After 12 months of field simulation).

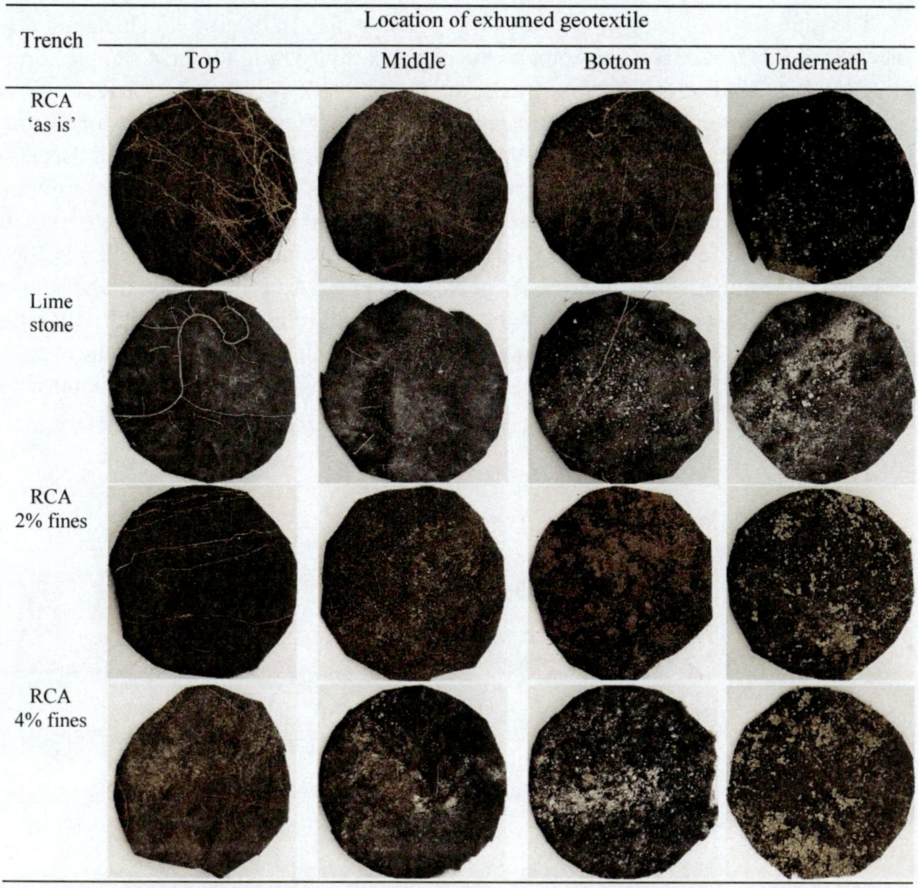

Fig. 21. Geotextile samples exhumed from French drains

Table 10. Permittivity values of unused and in situ nonwoven geotextiles

	Permittivity (sec^{-1})			
	Top	Middle	Bottom	Underneath
Unused geotextile	1.10*	1.10*	1.10*	1.10*
RCA 'as is'	0.98	0.80	0.75	0.59
Limestone	0.97	0.79	0.65	0.44
RCA with 2% fines	0.91	0.82	0.69	0.53
RCA with 4% fines	0.88	0.78	0.64	0.41

*permittivity for the unused geotextile regardless of location in French drain

The ratio for an unclogged geotextile is equal to 1.0, indicating no change in flow capacity of the geotextile. The comparison of permittivity ratio for each sample is presented in Fig. 22. For the exterior drains, the reduction of permittivity for RCA 'as is' is similar to limestone for the locations of Top and Middle; however, limestone shows larger reduction than RCA 'as is' in the locations of Bottom and Underneath. Because limestone has more fines (about 2.2%) than RCA 'as is' (about 0.8%), fines migrate downward and build up along the bottom portion of the French drain. The photos of the in-situ geotextiles (Bottom and Underneath) of limestone (see Fig. 21) clearly show more fines in comparison with exhumed geotextiles of RCA 'as is'. In case of the interior drains, both RCA with 2% and 4% fines exhibit overall reduction in permittivity ratio, but RCA 4% fines shows a larger reduction than RCA with 2% fines. Unlike the exterior drains, higher fines content (4% in this study) caused clogging along the all depths.

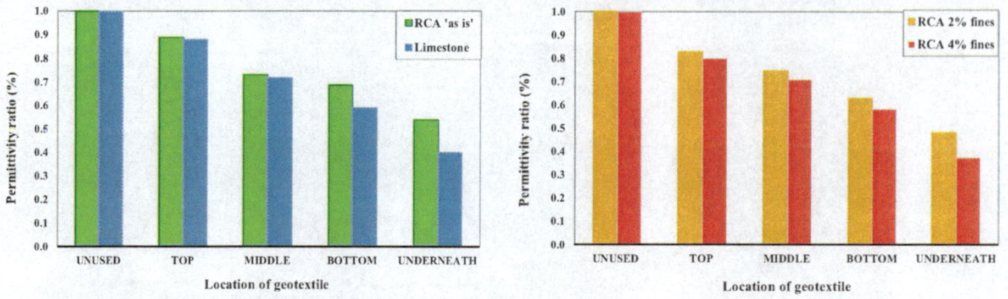

(a) Exterior drains (b) Interior drains

Fig. 22. Permittivity ratio of in situ geotextiles

5 Conclusions

This study was aimed at evaluating the effects of RCA on the long-term performance of French drain. The following conclusions have been made:

- Based on the field monitoring over 1 year, the drainage performance of RCA French drain is mainly controlled by in situ soil conditions (e.g. groundwater table, permeability of surrounding soils, etc.) and the amounts of excessive fines in the drain system; however, the aggregate type does not appear to be a major factor affecting the exfiltration drainage performance of French drain.
- When comparing RCA 'as is' (0.8% fines) with limestone (2.2% fines) it was observed that RCA performed better than limestone regarding flow rate (i.e. steady-state exfiltration rate at full storage) and maximum discharge rate (i.e. recovery period). To take into account the effect of the 1.4% fine difference between RCA 'as is' and limestone, the impact of fines between RCA 2% and RCA 4% was used. The 2% fine difference between RCA 2% and RCA 4% caused a reduction in flow rate of 17.8% based upon $Q_{avg\text{-}max}$. Even if RCA was extrapolated to contain 2.2% fines and was reduced the full amount assuming a 2% increase (even though it would actually be a 1.4%) this

would result in a $Q_{avg\text{-}max}$ value of 6.68(1–0.178) = 5.49, still above the $Q_{avg\text{-}max}$ for limestone of 5.38. So when taking into account the fine difference between RCA 'as is' and limestone French drains it can be seen that RCA performs similar if not better than limestone.

- The reduction of flow rate and discharge rate of French drain which consists of limestone is not only the material difference but also the fine content. An increase in fine content showed a noticeable decrease in flow rate. When comparing RCA 'as is' (0.8% fines) with limestone (2.2%), this 1.4% difference in fine content resulted in a 19.5% reduction in $Q_{avg\text{-}max}$. When comparing RCA 2% with RCA 4%, this 2% difference in fine content resulted in a 17.8% reduction in $Q_{avg\text{-}max}$. Thus, approximately a 1.5 to 2.0% increase in fine content within the range of 0% to 4% fines results in a 20% decrease in flow rate.
- The permittivity test results show that RCA with 4% fines show higher reduction in permittivity measurement after 12-month field conditioning. In addition, RCA with 2% fines and limestone drains have similar values of permittivity due to probably similar amount of fines content. It is concluded that the fines content is the major controlling factor on the permittivity measurement rather than the aggregate type.

References

1. Song, Y.H., Ooi, P.S.K., Hellebrand, E., Muenow, D.W.: Potential for Tufa precipitation from crushed concrete containing coarse basaltic and fine coralline sand aggregates. Environ. Eng. Geosci. **17**(1), 53–66 (2011)
2. Snyder, M., Bruinsma, J.: Review of studies concerning effects of unbound crushed concrete bases on pcc pavement drainage. Transport. Res. Rec.: J. Transport. Res. Board **1519**, 51–58 (1996)
3. Muethel, R.W.: Calcium Carbonate Precipitate from Crushed Concrete. Research Report No. R-1297. Michigan Department of Transportation, Materials and Technology Division, Lansing, MI (1989)
4. Tamirisa, R.: Study of highway base/subbase aggregates that cause depositions of calcareous "tufa" in Drains. Master's Thesis. University of Toledo Department of Civil Engineering, Toledo, OH (1993)
5. Nam, B.H., Behring, Z., Kim, J., Chopra, M.: Evaluate the use of reclaimed concrete aggregate in French drain applications. Technical Report: BDK 78 TWO 977–12, Florida Department of Transportation, Tallahassee (2014)
6. Nam, B.H., An, J., Youn, H.: Accelerated calcite precipitation (ACP) method for recycled concrete aggregate (RCA). Constr. Build. Mater. **125**, 749–756 (2016)

Application of Flowable Soil as Sustainable Backfill for Railway Track Stiffness Reinforcement at Bridge Transition Zone

Tack-Woo Lee[1], Tri Ho Minh Le[2], Dae-Wook Park[2](✉), and Jung-Woo Seo[2]

[1] Technology Research Division, Korea Rail Network Authority, Seoul, Korea
[2] Department of Civil and Environmental Engineering, Kunsan National University, Gunsan, Korea
dpark@kunsan.ac.kr

Abstract. In the railway track system, the abrupt change of track stiffness at the bridge transition zone leads to the faster degradation of track geometry, causing higher maintenance and rehabilitation costs. Many strategies have been introduced to solve this problem from geometry design to track stiffness equalization. Hence, this research focus on the development of new flowable backfill material for the railway bridge approach which has the combined effect of lightweight soil and strong resilient modulus. To cope with this objective, many laboratory experiments were conducted to determine the engineer properties of flowable soil from fresh to hardened stages. Then, the Finite Element (FE) model is integrated to simulate the behavior of flowable soil under different train speeds, passenger, and freight train types. The test results suggest that it is promising to use the flowable soil having foam as a proper backfill of the railway bridge approach since the low-weight properties help reduce the pressure to the below weak soil subgrade. The FE simulation also reveals that the settlement resistance of flowable soil is improved compared to the conventional cement-treated base. The high self-leveling characteristic of this material is expected to meet the strict requirement for wedge-shape application. In general, it is encouraged to use flowable soil as a sustainable backfill of the railway bridge approach zone .

Keywords: Flowable-fill · Track stiffness · Bridge approach · Bridge transition zone · Backfill · Settlement

1 Introduction

The world economy has been developed with a critical need in railway transportation. Recent findings report that around 6% of the annual income of the railway company will be used for the maintenance works (1, 2). The corresponded cost can be mitigated by apply sustainable construction methods, typically on locations for which the rail tamping technique is not efficient under long service life (3). The intensive dynamic loads generated by train-track interaction at bridge ends can impose a negative impact on railway integrity. A severe track settlement usually occurs at the transition area from

S. S. Kim et al. (Eds.): GeoChina 2021, SUCI, pp. 90–99, 2021.
https://doi.org/10.1007/978-3-030-80152-6_7

a bridge to the embankment. The rise in interaction force between the train and track caused by the differences in the track stiffness may lead to a faster demolition rate of track components (2, 3). The permanent settlement and vibration rate can also be considerably accelerated under high-speed trainloads. The vibration issue will impose a negative effect on the adjacent buildings and damage the ride comfort of track users (3). Therefore, a properly designed backfill at transition zone is expected to extend the service life of track structure based on the protection of concentrated dynamic force and excessive track settlement. Although there have been extensive efforts from railway specialists to upgrade the bridge approach zone using different strategies, it is yet to meet the desired performance, especially in the long-term scale. Besides, recent research has been focused on the design scale of the railway track system in the transition zone (2, 3). However, there are lack of findings on the backfill material of the bridge approach foundation associated with geotechnical property which also plays a very important role in the track stiffness support. Hence, this research aims to develop new backfill materials having a great rutting and vibration resistance, contributing to a sustainable bridge approach zone of the railway track system. Also, the high clay content soil is usually prohibited to be used as a backfill foundation at the bridge approach since the permanent resistance may easily occur. To cope with this objective, various mix design and test methods are used in this research. The local clay soil is utilized to fabricate backfill materials. Regards to the applicability characteristic of materials, the flowability and setting time tests are conducted in this research to figure out the fresh properties. Then, the compressive strength and freeze-thaw cycle tests are employed to evaluate the hardened behavior of backfill material. The optimal mix design generated from the experiment stage is applied in the numerical simulation to investigate the effectiveness compared to the conventional cement-treated base under high-speed railway loads.

2 Materials and Methods

2.1 Materials

The local soil at the bridge approach zone in this research comprises mainly of clay soil (Fig. 1a). The water content of excavated clay soil is 24.87% and the specific gravity is 2.677. The liquid limit, plastic limit, and plastic index (PI) is 32.6%, 23.8%, and 8.8%, respectively. This type of clay soil without treatment may impose a high risk of backfill settlement at the bridge approach zone. Hence, cement type II is used to improve the stiffness of local soil. In this research, the air foam is utilized to improve the workability of the slurry and reduce the mixed water. The air foam also provides the light-weight property for the backfill materials which help release the pressure for the whole track system and thereby, improve the track stiffness. The wet mixing method is applied in this research based on the suggestion from preliminary research and the author's team experience. In the wet mixing method, the dry components were first mixed with water before adding air foam. This step will protect the air foam cell from premature breakage due to impact with the dry soil or cement particles. Meanwhile, in the dry mixing process, the dry ingredients were pre-mixed, then air-foam and water were added to the combination. Regards to the wet mixing, cement, soil, and water (with superplasticizer) are first mixed for 3 min at 60 rpm. Then, the sufficient content of air foam is pumped into the slurry and the whole mixture is followed by an additional 3 min at the same rotation speed of around 60 rpm (Fig. 1b). The fresh slurry is cast into samples with a size of 70 mm in diameter and 140 mm in height for the unconfined compressive strength test, meanwhiles, the cylindrical mold size of the setting time test is 150 mm × 150 mm. In addition to the curing conditions, the samples are subjected to the cured chamber with a relative humidity of 100% and a temperature of 20 ± 2 °C until the testing day. The mix design is summarized in Table 1.

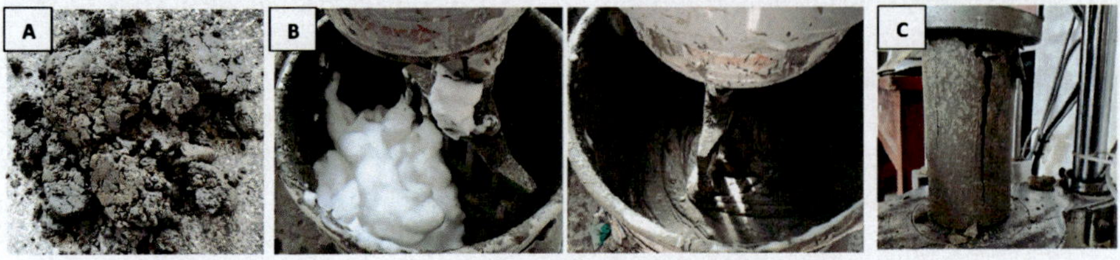

Fig. 1. (a) original soil; (b) mixing process; (c) unconfined compressive strength test.

Table 1. Mix Design (by cement weight)

	Mix	Cement (kg/m^3)	Soil (by weight of Cement)	Air-foam (% volume)	Water/Solid	Air-foam (% volume)
C100	C100–10	100	100%	10%	0.7	10%
	C100–20	100	100%	20%	0.65	20%
	C100–30	100	100%	30%	0.6	30%
C130	C130–10	130	100%	10%	0.7	10%
	C130–20	130	100%	20%	0.65	20%
	C130–30	130	100%	30%	0.6	30%
C160	C160–10	160	100%	10%	0.7	10%
	C160–20	160	100%	20%	0.65	20%
	C160–30	160	100%	30%	0.6	30%
C210	C210–10	210	100%	10%	0.75	10%
	C210–20	210	100%	20%	0.72	20%
	C210–30	210	100%	30%	0.68	30%

2.2 Experiment Test Methods

2.2.1 Flowability Test

The flowability is one of the most important traits of backfill materials. The flowability test is followed by ASTM D6103 (4). The mixture is confirmed to achieve the proper workability when the spread of fresh materials is within 200 ± 20 mm. The test results are measured from the average of three specimens.

2.2.2 Setting Test

Railway construction and maintenance works usually require strict time for traffic opening. Hence, it is very necessary to ensure the fast hardening of backfill material for other railway construction steps. The setting test is conducted based on the standard ASTM C403 (5). Related research on flowable fill materials suggests that it should obtain the initial setting strength of around 2.74 MPa for approximately 24 h. Three samples are used to calculate the average setting time value.

2.2.3 Unconfined Compressive Strength Test (UCS Test)

The fresh samples were cured in the 100% humidity condition at room temperature of around 23 °C. After 24 h, the samples were stripped out from the molds and they were let to be cured in the same condition until the testing day (28th day). In this research, the UCS value is expected to vary from 1 to 1.4 MPa which ensures the stress-bearing as well as excavation ability. The UCS test is conducted in accordance with the ASTM D 4832 (6). The universal testing machine with a loading rate of 1 mm/min is applied on three replicates to measure the compressive strength of one condition (Fig. 1c).

2.3 Numerical Simulation

In this study, the commercial program Abaqus is used to develop the numerical simulation of the railway bridge approach zone under train loads. The best mix design in the experiment stage is applied in the backfill location of bridge transition to compare with the traditional cement-treated base. The main concept of the simulation is to generate the actual moving trainloads from the bridge zone to the backfill zone. Then, the track stiffness differences are evaluated from each location. Based on the suggestions from Korea Railroad Design (7, 8), the track stiffness ratio is defined as track stiffness of bridge zone (k1) by the track stiffness at approach zone (k2). To guarantee the safety of the railway track system, the track stiffness ratio should be controlled at lower than 2.5 and 1.8 for the normal speed and high-speed trains, respectively. Also, A 10 Hz positive full-sine waveform having a peak load of 131.31 kN and a min load of 30 kN is used to generate the cyclic load to the backfill foundation. This design aims to evaluate the dynamic performance of trains under repeated load. The 2D FEM is designed with the 4-node bilinear plane strain quadrilateral element type (CPE4). The bottom of the model is fully constrained to all displacement, meanwhiles, the other part is allowed to move vertically. Regards to the interface between the wheel and rail, the frictionless and tangential properties are defined. The remained interfaces are defined as tie contact to simplify the simulation process.

2.3.1 Loading Condition

The concentrated force from the train wheel load is designed to achieve the central force of around 131.31 kN. This value represented the dynamic wheel load generated by the KRL-2012 Passenger train type operated at the speed of 300 km/h (7, 8). The following equation is used to calculate the load value:

$$\boldsymbol{P_{dyn} = P_{eff} \times DAF} \tag{1}$$

Where,

- 60 < V ≤ 350 km/h passenger train, $\boldsymbol{DAF = 1 + t \times Ø\left(1 + 0.5\frac{V-60}{190}\right)}$

Where,

- $\boldsymbol{P_{st}}$: Static wheel load (kN); $\boldsymbol{P_{eff}}$: Effective wheel load ($\boldsymbol{P_{st}} \times$ 1.2, kN)
- $\boldsymbol{P_{dyn}}$: Dynamic wheel load (kN); DAF: Dynamic Amplitude Factor
- t: Weight of standard deviation depending on a confidence interval of probability (t = 1 applied to roadbed calculation).
- Ø: Coefficient dependent on track quality (Ø = 0.2 for good tracks)

The schematic illustration of the finite element model used in this study is portraited in the Fig. 2.

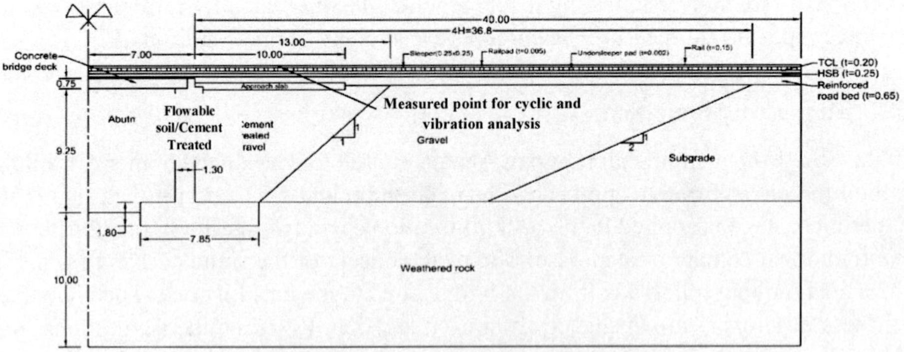

Fig. 2. Bridge approach zone model.

2.3.2 Material Inputs

The materials inputs used in this research are based on Korea Railroad Design and preliminary research (1, 7, 8). It should be noted that the constitutive model and elastoplastic material followed by Drucker-Prager/Cap yield method are applied to precisely simulate the actual dynamic response of the whole track foundation. From the guidance of Korea Railroad Design, the friction angle, dilation angle, and cohesion C of the ballast layer are designed at 39, 9, and 1, respectively.

3 Results and Discussions

3.1 Experiment Results

3.1.1 Flowability Test

The workability of the flowable fill material is investigated through the flowability test shown in Fig. 3. The initial flowability test results indicate that clay soil mixture requires significant high water content to achieve the ideal flowability. This may be due to the high absorption of clay particles in the mixture. The trapped water in the clay slurry may impose a negative effect on the backfill materials, thus, air foam and superplasticizer were employed in this study to solve the issue. Within constant water/solid ratio, the test results indicate that mixture having higher cement content also showed unsatisfied workability due to the fine particle size of cement and the water adsorption ability. It is reasonable to find that the higher the air foam levels, the higher the flowability value. However, the air foam should not be overused to avoid the critical subsidence problem and the bleeding phenomenon due to the breakage of air foam. In this research, the mixtures with different air foam ratios shown in Table 1 were modified to meet the optimum flowability of around 200 mm. This modification process mainly depends on the initial mixing water. For examples, the water content was minimized in the high air-foam mixtures, meanwhile, higher volume of fluid was added into the low air-foam condition to improve the workability. After the water/solid ratio modification process, all mixtures achieve the desired flowability of around 200 mm which can be properly applied in practice.

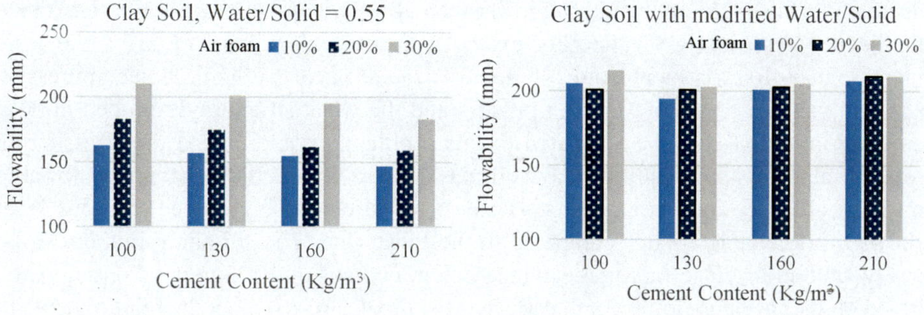

Fig. 3. Flowability test results

3.1.2 Setting Test

The setting time test results are illustrated in Fig. 4. Cement content helps flowable fill mixture accelerates the hardening process noticeably. The hydration process of backfill material at the railway bridge approach should trigger at a fast rate to ensure the followed construction step. Among all mixture designs, the mixture having cement content of 210 kg/m^3 and air foam level of 10% reach the fastest setting time at around 15 h which outperforms the remained mixtures. Meanwhiles, the other mix designs achieve the initial setting time longer than 1 day which may not properly be applied in actual railway construction. The setting time test also reveals that appropriate foam content should be greatly considered since this component may prolong the hardening process of flowable material.

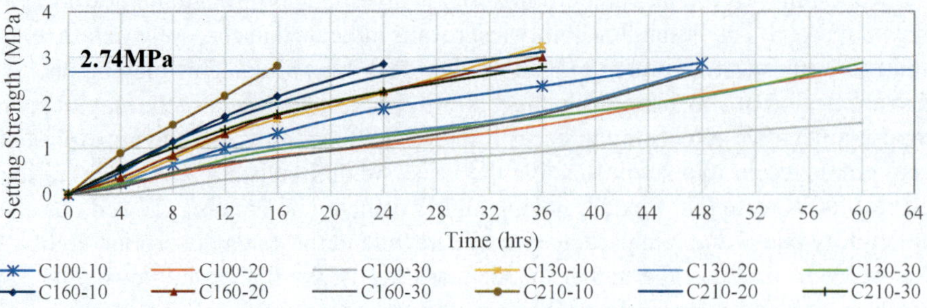

Fig. 4. Setting time test results.

3.1.3 28 Days Unconfined Compressive Strength Test

The unconfined compressive strength test results are exhibited in Fig. 5. Overall, the UCS recored in this flowable materials ranging from 0.05 to 1.3 MPa. As regards the first, the UCS results agree with the findings from the above setting time test since mixtures having 210 kg/m^3 dominates the remained mix designs. It is suggested that the air foam level should not be used at lower than 20% by volume of total mix to exclude the sudden drop in UCS value. For example, at the cement level of 210 kg/m^3, the increase in air foam content from 20% to 30% lead to a nearly 60% drop in compressive strength value. The slow hydration process and the weak structure developed in 30% air foam mixtures may be attributed to this problem. In addition to the stress-strain relationship, the application of clay soil may lead to the ductile behavior of flowable material after reaching peak stress. Besides, mixture with higher cementitious level shows brittle characteristics compared to the lower one. The softening failure can be easily spotted in those mixtures having cement content of 100 kg/m^3. Consequently, based on the investigation from the above tests, mix C210–10 meet the requirements to be applied in the simulation process. From the stress-strain curve analysis, the elastic modulus of the optimal mixture is calculated and this value is approximately 300 MPa.

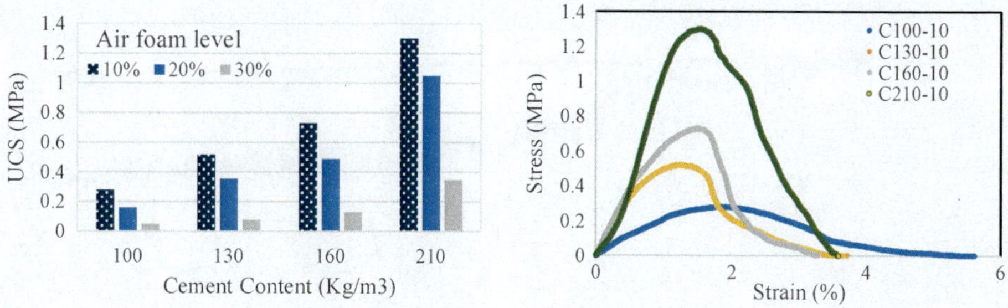

Fig. 5. Unconfined compressive strength

3.2 Simulation Results

The simulation investigation is shown in Fig. 6, 7, and 8. The FEM results indicate that it is very promising to use flowable soil as a backfill material of the railway bridge approach to improve the durability of the whole track structure. Although the wedge shape design of backfill provides a smooth reduction in track stiffness, under moving train loads, it can be seen that flowable soil remarkably improve the track settlement resistance at weak rail spot compared to the conventional cement-treated base (Fig. 6). The effectiveness is more prominent since the concrete slab track is designed. By measuring the k1/k2 ratio in concrete slab track, the value of flowable soil and conventional CTB is approximately 1.6 and 2, respectively. The settlement gap between the two railway track options is around 0.5 mm which may be critical under the long-time service of high-speed trains. Hence, this finding agrees with recent research about the efficiency of concrete slab track.

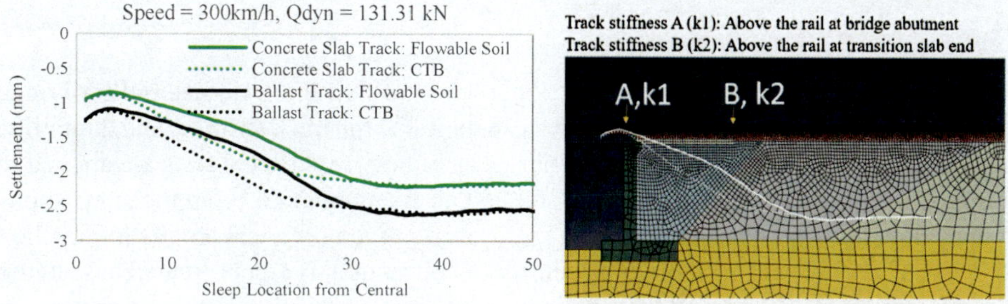

Fig. 6. The settlement results caused by moving train at the railway bridge approach

The benefit of flowable soil is also verified by conducting repeated trainload (10 Hz). It can be observed from Fig. 7 that the settlement value of flowable soil develops gradually from 0 to 2 mm, meanwhiles, the conventional cement-treated base increase significantly which may lead to the fast degradation of the whole track system. The novel advantage of flowable fill is contributed by its high elastic modulus of around 300 MPa and the light-weight property of 1400 kg/m^3. In contrast, these values of cement-treated base are 120 MPa and 2000 kg/m^3, respectively. The proficiency of flowable soil is also confirmed through the vibration under train loads (Fig. 8), the simulation results indicate that the

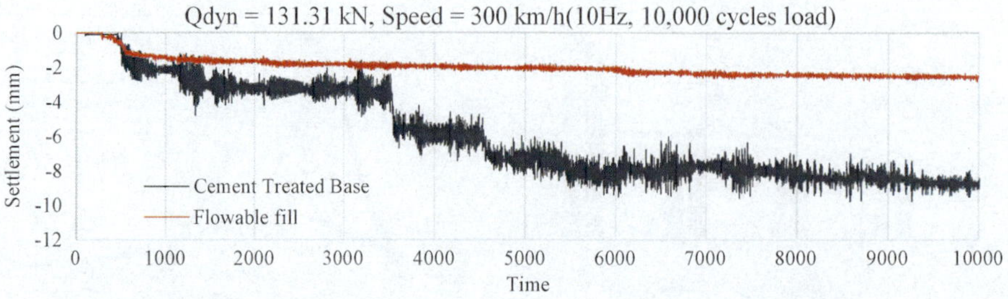

Fig. 7. The permanent settlement results of backfill bridge approach.

acceleration of flowable soil is twice lower than the traditional method which will help enhance the durability of track structure and the comfort of track users.

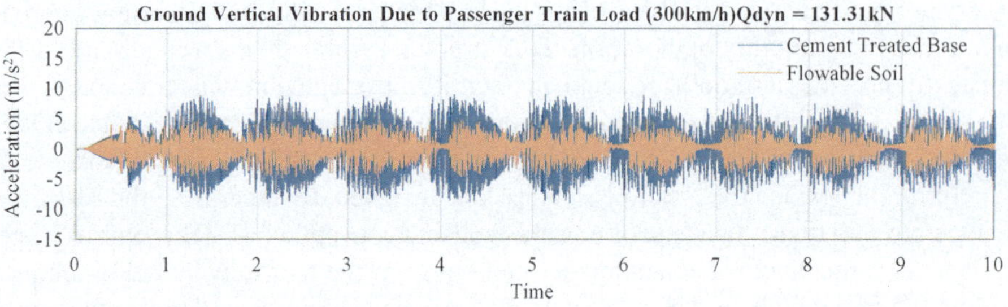

Fig. 8. The vibration results in the backfill transition zone.

4 Conclusion

The main aim of the research is to develop a practical flowable soil material for bridge transition zone of railway structure which can reinforce the track stiffness and durability. Hence, to cope with this objective, a series of laboratory experiment tests are applied in flowable soil material using local clay soil and air foam method. From the suggestions from test results, the optimum mix design is used in the numerical simulation method to evaluate its effectiveness as a replacement of conventional cement-treated base under actual train loads. The laboratory tests indicated that clay soil has a very high absorption ability which may lead to the high mixing water usage. Air foam level help improves the workability of flowable soil mixture; however, it should be properly used to avoid the prolonged effect in setting time and poor compressive strength gain. The numerical simulation on the dynamic response of flowable soil under high-speed trains reveals that the new flowable material not only resolves the sudden drop in track stiffness at bridge approach transition zone but also acquire high permanent settlement resistance under repeated train loads.

Acknowledgment. This research was supported by a grant from R&D Program of the Korea Railroad Network Authority, Republic of Korea.

References

1. Ognibene, G., Powrie, W., Le, P. L., Harkness J.: Analysis of a Bridge Approach: Long-Term Behavior from Short-Term Response. Railway Engineering, Edinburgh, UK (2019)
2. Eric B.: Railway Track Stiffness. Dynamic measurements and evaluation for efficient maintenance. Doctoral Thesis. Royal Institute of Technology (KTH) (2009), ISSN 1651–7660
3. Haoyu W., Mika S., Valeri M., Bruce W. Analysis of the Dynamic Wheel Loads in Railway Transition Zones Considering the Moisture Condition of the Ballast and Subballast. MDPI: Applied Sciences (2017)
4. ASTM D 6103. Standard Test Method for Flow Consistency of Controlled Low Strength Material. ASTM International, West Conshohocken, PA (1997)
5. ASTM C 403. Standard Test Method for Time of Setting of Concrete Mixtures by Penetration Resistance. *ASTM International*, West Conshohocken, PA (1999)
6. ASTM D4832. Standard Test Method for Preparation and Testing of Controlled Low Strength Material (CLSM) Test Cylinders. *ASTM International*, West Conshohocken, PA
7. Korean Railroad Research Institute. https://www.krri.re.kr/html/en/. Accessed August 2020
8. Tri H. M. Le., Seong-Hyeok L., Dae-Wook P.: Evaluation on full-scale testbed performance of cement asphalt mortar for ballasted track stabilization. Construc Building Mater. **254** (2020)

Numerical Modeling of Grout Injection for Hybrid Bored Prestressed Concrete Cased Piles

Hesong Hu[1], Sudheer Prabhu[2], Xiaobin Chen[3], and Tong Qiu[4](✉)

[1] Guangzhou Institute of Building Science Co., Ltd., 833 North Baiyun Ave, Guangzhou 510440, China
hhs@gibs.com.cn

[2] Department of Civil and Environmental Engineering, Pennsylvania State University, University Park, PA 16802, USA
ssp23@psu.edu

[3] School of Civil Engineering, Central South University, 22 South Shaoshan Rd, Changsha 410075, Hunan, China
chen_xiaobin@csu.edu.cn

[4] Department of Civil and Environmental Engineering, Pennsylvania State University, University Park, PA 16802, USA
tuq1@psu.edu

Abstract. In this paper, a numerical analysis of grout flow in a narrow 10-mm soil-pile gap was conducted using the ANSYS Fluent software. The Volume of Flow (VOF) method was used to model the grout flow through the soil-pile gap. Two-phase transient flow simulations were carried out with grout and air being the two immiscible fluids. Four different pile diameters of 800, 1000, 120,0 and 1400 mm were considered and for each pile diameter, grouts with two different water-cement ratios of 0.4 and 0.5 were pumped into the soil-pile gap. Numerical results indicate that: (1) for the same pile diameter and grout pressure, the height of grout flow increases as the water-cement ratio increases from 0.4 to 0.5; (2) for the same grout pressure, the height of grout flow decreases as the pile diameter increases; (3) for a give pile diameter and water-cement ratio, the final grout height is mainly dependent on the grout pressure but insensitive to the inlet configuration; and (4) the two-inlet configuration is more effective in eliminating entrapped air at the bottom of the pile than the three- and four-inlet configurations.

Keywords: Grouting · Non-Newtonian fluids · Two-phase transient flow · Viscosity

1 Introduction

Prestressed high-strength concrete (PHC) piles have gained increasing popularity from the construction industry due to their high bearing capacity, reliable quality, and cost advantages (Aoki 1998; Dockray 2001; Bradshaw and Baxter 2006; Nie et al. 2016). For the installation of large-diameter (D $\geq$ 800 mm) PHC piles, they are lowered into the

S. S. Kim et al. (Eds.): GeoChina 2021, SUCI, pp. 100–112, 2021.
https://doi.org/10.1007/978-3-030-80152-6_8

ground simultaneously with the continuous boring of the hole without excessive noise, vibration, and damage of the pile tip. Chen et al. (2018) developed a novel installation technique for large-diameter PHC piles in stiff strata, where the bore hole is slightly larger than the pile diameter to facilitate the lowering of the pile into the ground and the soil-pile gap is subsequently pressure-grouted to increase the bearing capacity. Figure 1 shows the hardened grout after removing the surrounding soil. Through full-scale load tests, Chen et al. (2018) demonstrated that pressure grouting at soil-pile gap drastically increases the bearing capacity. However, the flow of grout under pressure through the narrow soil-pile gap, approximately 10 mm in width, is relatively unknown. A non-uniform distribution of grout at the soil-pile gap along the pile may negatively impact the bearing capacity. Hence, more research is needed to study the grout flow in the narrow soil-pile gap for large-diameter PHC piles.

Fig. 1. Pile head after removing surrounding soil showing hardened grout

In this paper, a numerical analysis of grout flow in a narrow 10-mm soil-pile gap was conducted using the ANSYS Fluent software. The Volume of Flow (VOF) method was used to model the two-phase transient grout flow through the soil-pile gap with grout and air being the two fluids. Four different pile diameters of 800, 1000, 1200, and 1400 mm were considered and for each pile diameter, grouts with two different water-cement ratios of 0.4 and 0.5 were pumped into the soil-pile gap. In this study, the surrounding soil is modeled as a frictional boundary condition and its potential deformation is not considered for simplicity. In the following sections, the development of the numerical model was first presented, followed by detailed discussions on the effects of water–cement ratio, pile diameter, inlet size, and number of inlets and grout pressure on the grout flow.

2 Numerical Model

For the cylindrical piles considered, grout was injected through two diametrically opposite points on the pile. These points were located 1 m above the bottom of the pile. Figure 2 shows the schematic diagram of the problem considered. A uniform gap of

10 mm was assumed between the pile and the ground. Numerical simulations of grout flow in the gap between the concrete pile and soil were carried out in a commercial software ANSYS Fluent Version 19.1. In the numerical simulations, taking advantage of symmetry, only a quarter of the problem was modelled. In the field, 30 mm diameter tubes are used to inject grout. However, to reduce the complexity of the mesh, a square cross section of inlet was considered with equivalent area as that of the circle. To improve the efficiency and reduce the computation cost, only 15 m of the soil-pile gap was modelled. Four different pile diameters of 800, 1000, 1200 and 1400 mm were considered and for each pile diameter, grouts with two different water-cement ratios of 0.4 and 0.5 were pumped into the soil-pile gap. For all the geometries considered, mesh sensitivity analysis was carried out to determine an appropriate mesh with the optimum efficiency and accuracy. For example, Fig. 2 shows the mesh used for the 800-mm diameter pile. Each element has an average size of 0.01 m. The mesh generated for the soil-pile gap for the 800-mm diameter pile consists of 198132 nodes while the other meshes with pile diameters of 1000, 1200, and 1400 mm have 246164, 294196, and 339226 nodes, respectively.

In the current study, the Volume of Flow (VOF) method was used to model the grout flow through the soil-pile gap. In general, the VOF method is used to study the flow of two or more immiscible fluids and their interface tracking. A single set of momentum equations are solved and the volume fraction of each fluid is tracked throughout the domain. Values of the solved quantities are calculated at the centroids and it is assumed that the quantities vary linearly across a cell (Fluent Manual 2012). The VOF method is Eulerian method that uses volume fraction indicator to determine the interfaces of different fluids in the computational domain (Liovic et al. 2000). Rather than tracking the interface, the volume of each material in each cell is evolved in time and the interface is reconstructed from the values of volume at current timestep (Pillod and Puckett 2004). Due to its simplicity and computational efficiency, the VOF method has been widely used by various researchers (Renardy et al. 2002; Kunkelman and Stephan 2009). A two-phase transient flow simulation was carried out with grout and air being the two immiscible fluids. Each simulation was carried out for 25 s in real time and the height reached by the interface between the two fluids was monitored. The grout pressure was assumed to be in the range of 300 kPa - 500 kPa. Initially, the soil-pile gap was assumed to be completely filled with air and the typical properties of air was utilized ($\rho = 1.225\,\text{kg/m}^3$; viscosity 1.79×10^{-5} Pa.s).

Different models have been developed to simulate the flow behavior of grout and fresh concrete. Some of these models are Newtonian fluid model, Power law model, and Bingham fluid models. Eriksson et al. (2000) modelled grouts as Newtonian and Bingham fluids to understand the grout spreading while Lachemi et al. (2003) used Bingham model to approximate the rheological flow properties of cement paste. However, Tattersall and Banfill (1983) showed that grouts are viscoplastic fluids and flow occurs only beyond a certain yield stress. The power law model is a popular model to describe the flow of different fluids; however, it is only suitable to capture fluids having zero yield stress. Rosquoet et al. (2003) showed that cement paste acted as a shear thinning fluid as the viscosity decreased with increase in strain rate; while, Cyr et al. (2000) showed that cement pastes added with superplasticizers exhibited a shear thickening behavior

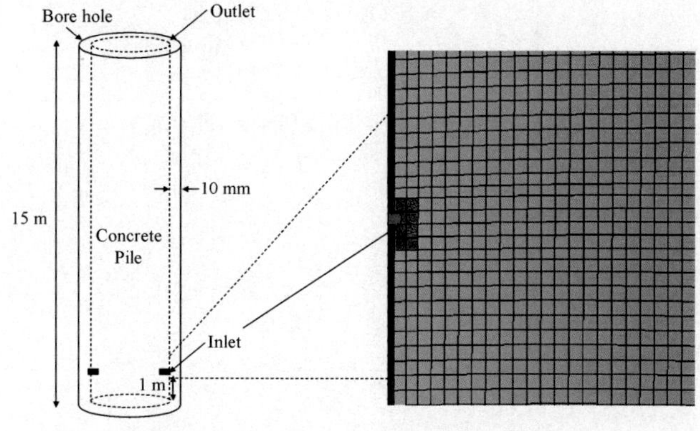

Fig. 2. Problem geometry and mesh: (a) schematic diagram of the two-inlet configuration; (b) mesh generated near the inlet

as the viscosity increased with increase in strain rate. None of the models mentioned above could account for all of these observed flow behaviors in cement paste or grout. Larrard et al. (1997) noted that modelling fresh concrete as Bingham fluid could lead to negative values of yield stress. Atzeni et al. (1985) compared different rheological models including Newtonian, Bingham, power equation and Herschel – Bulkley (HB) models in capturing the flow behavior of cement paste; it was noted that the HB model captures the flow behavior accurately and similar results were also noted by Larrard et al. (1997). In Bingham fluids, flow occurs once the yield stress is exceeded; while in the HB model, flow occurs even below the yield stress with a much higher viscosity compared to the actual viscosity of the fluid thus approximately providing a rigid behavior below the yield stress (see Fig. 3).

The HB model is divided into two regimes: one below the critical strain rate ($\dot{\gamma_c}$) and one above it. The viscosity of fluid below the critical strain rate is given by Eq. 1 while the viscosity of the fluid above $\dot{\gamma}_c$ is given by Eq. 2. The power term n decides whether the fluid acts as shear thickening ($n > 1$), shear thinning ($n < 1$), or Bingham ($n = 1$) fluid. Figure 3 shows the variation of shear stress with strain rate for different fluids considered. A very small value of $\dot{\gamma}_c$ can capture an almost 'rigid' behavior for strain rates below $\dot{\gamma}_c$. For $\dot{\gamma} < \dot{\gamma}_c$

$$\eta \frac{\tau_o}{\dot{\gamma}} + K(\frac{\dot{\gamma}}{\dot{\gamma}_c})^{n-1} \tag{1}$$

and for $\dot{\gamma} > \dot{\gamma}_c$

$$\eta \frac{\tau_o(2 - \gamma/\dot{\gamma}_c)}{\dot{\gamma}_c} + K[(2 - n) + (n - 1)\left(\frac{\dot{\gamma}}{\dot{\gamma}_c}\right)] \tag{2}$$

where K is the consistency which is defined as the apparent viscosity of the fluid at a shear rate of 1 s^{-1}; η is the viscosity; τ_o is the yield stress of the fluid; and $\dot{\gamma}$ is the strain rate. The values of K, τ_o, n and $\dot{\gamma}_c$ are calculated by curve fitting the experimental shear stress versus strain rate obtained using marsh cone or viscometer.

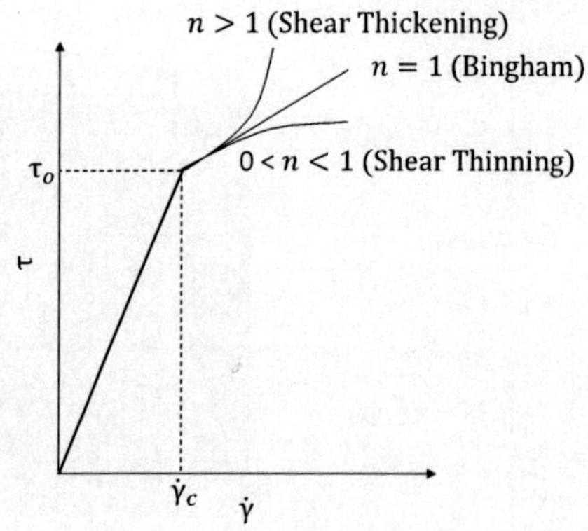

Fig. 3. Variation of shear stress with strain rate for HB model

The grout was injected into the gap through the inlets as shown in Fig. 2 and was modelled using the HB model. Nguyen et al. (2006) carried out Marsh cone tests to understand the rheological properties of grouts and observed that for the water-cement ratios of 0.4 and 0.5, the grout behaved as shear thinning fluid. The study also derived the HB model parameters for grout at these water-cement ratios and are shown in Table 1. A small value (0.01) was provided for the critical shear strain rate ($\dot{\gamma}_c$) which provides a high initial viscosity to the flow below the yield stress of the grout.

Table 1. Properties of the grout considered (Nguyen et al. 2006)

Water-cement ratio	ρ (kg/m^3)	Yield stress τ_o (Pa)	n	Consistency K (Pa.s^n)	Critical strain rate $\dot{\gamma}_c$ (s^{-1})
0.4	1931	15.7	0.390	14.28	0.01
0.5	1816	8.1	0.408	3.87	0.01

For grouting under high pressure, the grout flow might be turbulent (Kishi et al. 2014) and is modelled using a $k - \omega$ turbulence model, which is a two-equation model with two extra transport equations to represent turbulent properties of flow. The Wilcox $k - \omega$ model used in the current study has been calibrated and validated by various researchers (Menter et al. 2003; Chan et al. 2012) and has been shown to be applicable in wall-bounded flow problems. The model attempts to predict turbulence by two partial differential equations for kinetic energy (k) and specific dissipation rate (ω), given by:

$$\frac{\partial(\rho k)}{\partial t} + \frac{\partial(\rho U_j k)}{\partial x_j} = \frac{\partial}{\partial x_j}\left[\left(\mu + \frac{\mu_t}{\sigma_k}\right)\frac{\partial y}{\partial x_j}\right] + P_k - \beta' \rho k \omega - P_{kb} \quad (3)$$

$$\frac{\partial(\rho \omega)}{\partial t} + \frac{\partial(\rho U_j \omega)}{\partial x_j} = \frac{\partial}{\partial x_j}\left[\left(\mu + \frac{\mu_t}{\sigma_k}\right)\frac{\partial y}{\partial x_j}\right] + \alpha \frac{\omega}{k} P_k - \beta \rho \omega^2 + P_{\omega b} \quad (4)$$

where ρ is the density; U_j is the velocity vector in the jth direction; ω is the specific turbulence rate; μ is the fluid viscosity; P_k is the production rate of turbulence; β', β, α,

σ_k, and σ_ω are model constants; $P_{\omega b}$ and P_{kb} are the buoyancy terms; μ_t is the turbulent viscosity; and σ_k and σ_ω are turbulent Prandtl numbers. The model is a closure for the Reynolds averaged Navier-Stokes equations. The detailed description of these variables and their default values have been described in detail in Fluent Manual (2012) and the detailed description of the model is provided in Wilcox (2006). The cylindrical inner and outer walls as well as the side walls were modelled as stationary walls with the non-slip boundary conditions.

The wall roughness reduces the mean velocity of the grout near the wall and depends heavily on the roughness characteristics (Kays and Crawford 1993; Schlichting 1979). To consider the roughness of the bored hole and concrete pile on the grout flow, the term equivalent sand roughness is used. The rough surface is replaced by sands with uniform or non-uniform diameters such that the surface replicates a sand paper surface. The equivalent sand grain roughness depends on the pattern of the roughness on the surface, density and shape of the groves on the surface. In Ansys fluent, roughness of the walls is implemented using the term equivalent sand grain surface roughness coefficient (K_s) and roughness constant (C_s). Moody (1944) noted that for concrete surfaces the value of K_s varied from 0.3 mm (smooth) to 3 mm (rough) depending on the surface roughness. In all the simulations considered in the current study, the highest value of K_s (3 mm) was used which implies that the concrete surface and bored hole were very rough. The roughness of the walls was assumed to be uniform and thus a value of 0.5 was used for the roughness constant (C_s).

3 Results and Discussions

3.1 Effect of Water–cement Ratio and Pile Diameter on the Height of Grout Flow

The multiphase transient analysis was carried out to understand the height of grout flow in the soil-pile gap for different pile diameters and grouts with varying water-cement ratio. Typical inlet size of 30 mm in diameter was used in all the simulations in the current section and the grout pressure was kept constant at 500 kPa. The diameter of the pile considered was 800 mm and the water–cement ratio was 0.4. Figure 4 shows the volume of grout filling the soil-pile gap at different time intervals. The grout-air interface was uneven at the top and thus, the average height of the grout-air interface from the bottom of the pile was taken as the height of grout flow. Figure 5 shows the corresponding pressure contour plots developed at the same time intervals. Initially, throughout the soil-plie gap the pressure was negligible as the volume was completely filled with air and there was a rapid expansion of the injected grout (see Fig. 5(a)). Subsequently, more volume of the soil-pile gap was filled by the grout which led to the development of pressure contours. Figure 5(d) shows that at later stages, the pressure distribution at the top (away from the inlets) had a constant gradient with depth (i.e., unit weight of grout); however, the pressure distribution at the bottom was concentrated with high pressures in the vicinity of the inlets.

Figure 6 shows the height of grout flow for different pile diameters and water–cement ratios. It can be observed from Fig. 6 that for the same pile diameter, the height of grout flow increases as the water-cement ratio increases from 0.4 to 0.5. This can be attributed to the decrease in viscosity or resistance to grout flow with an increase in water–cement ratio. Another key observation is that as the pile diameter increases, higher weight of grout needs to be pumped up and hence, for the same grout pressure the height of grout flow decreases. For a pile diameter of 800 mm the grout with water-cement ratio 0.5 reaches a height of 12.3 m after 25 s while for 1400 mm diameter pile, the grout only reaches a height of 9.9 m. It is also evident from Fig. 6 that water–cement ratio of the grout plays a much more significant role in the height of grout flow compared to the pile diameter. For the same pile diameter of 800 mm, an increase in water-cement ratio from 0.4 to 0.5 increases the height of grout flow by 50%. In all the simulations considered, the height of grout flow increases rapidly with time initially, but the rate of increase reduces with time. As the simulations were computationally expensive and the increase in height of grout flow was marginal after 25 s, particularly for water-cement ratio of 0.4, simulations were terminated after 25 s. Thus, the height of grout flow after 25 s was taken as the ultimate height reached by the grout for simplicity.

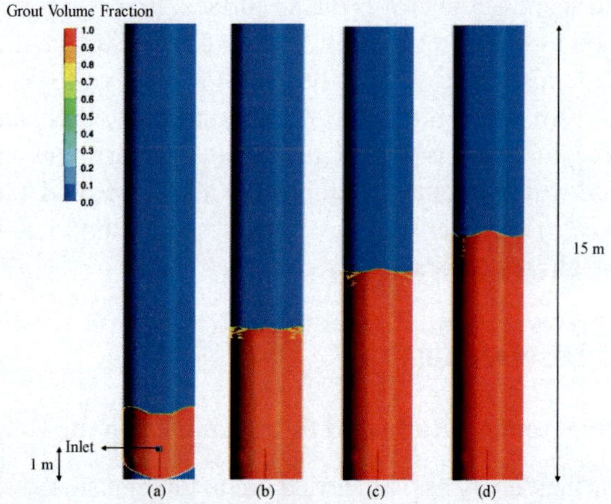

Fig. 4. Grout volume fraction at different times for 500 kPa grout pressure and 0.4 water-cement ratio: (a) 2 s; (b) 7 s; (c) 15 s; (d) 25 s (not to scale)

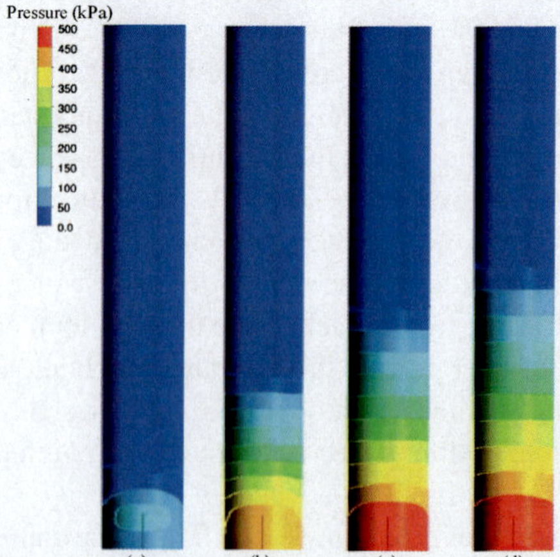

Fig. 5. Pressure contours at different times for 500 kPa grout pressure and 0.4 water-cement ratio: (a) 2 s; (b) 7 s; (c) 15 s; (d) 25 s (not to scale)

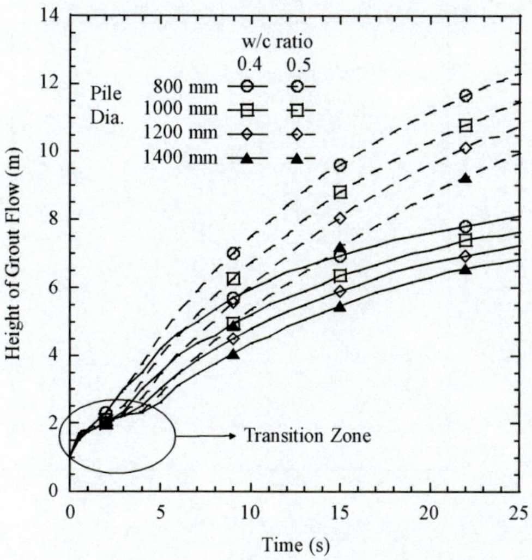

Fig. 6. Height of grout flow vs. time for different pile diameters and water-cement ratios

It can also be observed from Figs. 4 and 6 that initially, the grout injected into the soil-pile gap is spread to fill the bottom 1-m of space. At this stage, the increase in height of grout flow is minimal and this is termed as the transition zone. Once the bottom space is completely filled with grout, the subsequent injection of grout is directly pushing the grout to flow higher.

3.2 Effect of Inlet Size on the Height of Grout Flow

The effect of inlet size on the height of grout flow was studied by comparing the results from inlet diameters of 20 mm and 30 mm, where the pile diameter was 800 mm and the water-cement ratio was 0.4. Figure 7 shows that initially, the height of grout flow is much higher in the case of 30 mm diameter inlet. This is justified as the volumetric flow rate of the grout from the 30 mm diameter inlet would be higher than that from the 20 mm diameter inlet. Due to the high flow rate in the case of 30 mm diameter inlet, the transition time is lower compared to the case of 20 mm diameter inlet as evident from Fig. 7. With time, the gap between the two curves (Fig. 7) is reduced and if ample time was provided, the final height of grout flow will be relatively insensitive to the inlet size. This can be explained using Fig. 7 which shows the pressure at the bottom of the pile being equal to the grout pressure in all directions when the grout reaches the final height (see Fig. 5(d)). The final height of grout is supported by the grout pressure acting on the annular area of the soil-pile gap. Hence, the final height is mainly dependent on the grout pressure and insensitive to the inlet size.

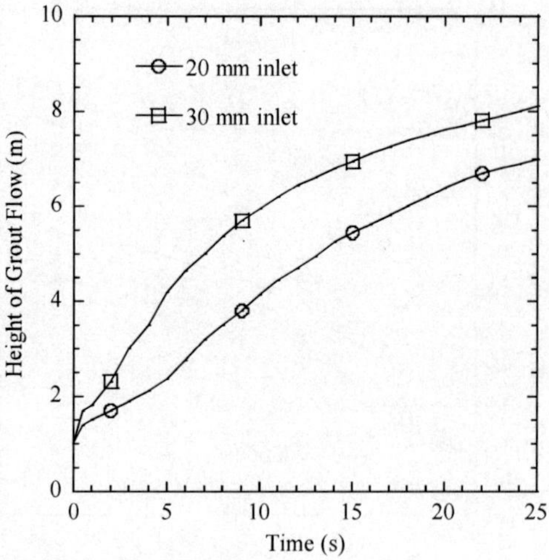

Fig. 7. Height of grout flow vs. time for different inlet sizes for 800 mm pile diameter, 0.4 water–cement ratio, and 500 kPa grout pressure

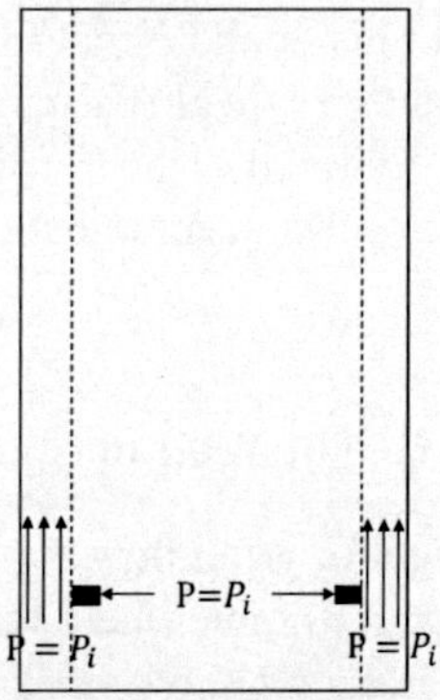

Fig. 8. Grout pressure supporting final height of grout

3.3 Effect of Number of Inlets and Grout Pressure on the Height of Grout Flow

In this section, the effect of number of inlets and the grout pressure on the height of grout flow was studied. The number of inlets to inject the grout was varied as 2, 3 and 4 and for each of these configurations, two grout pressures of 300 kPa and 500 kPa were applied. Application of pressures beyond 500 kPa made the flow unstable and led to intermixing of the two phases, which could not be simulated using the model considered. Therefore, the maximum pressure in this study is limited to 500 kPa. For the two-inlet configuration, only a quarter of the problem was modelled as discussed in the previous section. Similarly, 1/6th and 1/8th of the problem was modeled for the three-inlet and four-inlet configurations, respectively. Figure 9 shows the schematic diagram of the different configurations modelled in the current study. The corresponding number of nodes in the mesh for these configurations was 198132 (two-inlet), 135090 (two-inlet) and 99066 (four-inlet), respectively. In all the cases considered in the current section, the water–cement ratio was taken to be 0.4 and the pile diameter was 800 mm. Figure 10 shows the height of grout flow for different inlet configurations and grout pressures. It can be observed from Fig. 10 that both inlet configuration and grout pressure have a direct influence on the height of grout flow. As the number of inlets increases, the height of grout flow also increases for a given grout pressure. For the same inlet configuration, as the grout pressure increases, the height of grout flow increases.

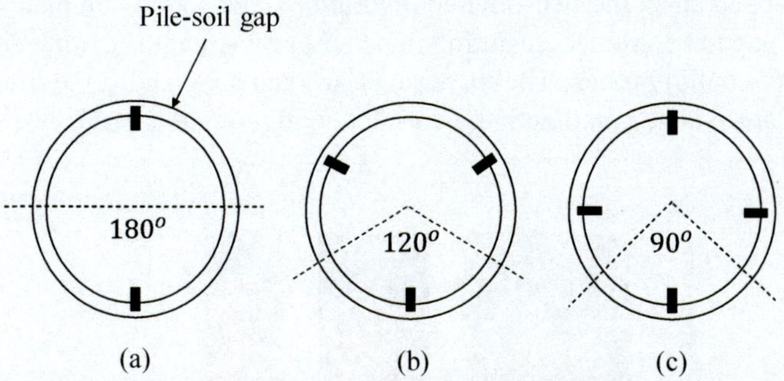

Fig. 9. Different inlet configurations: (a) two-inlet; (b) three-inlet; (c) four-inlet

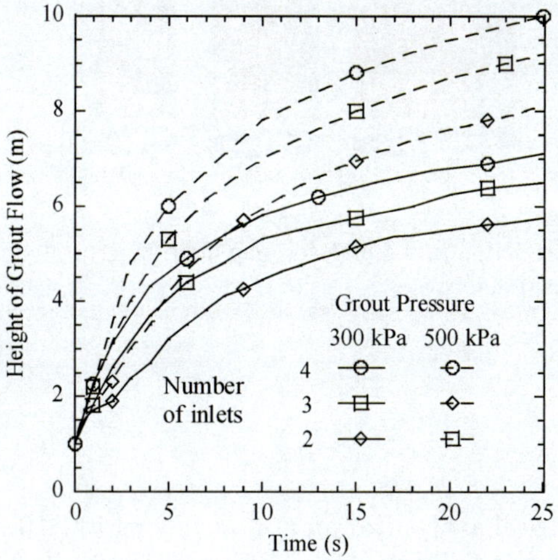

Fig. 10. Height of grout flow vs. time for different inlet configurations and grout pressures

It is observed from Fig. 10 that as the number of inlets increases, the height of grout flow also increases. However, increasing the number of inlets might not be advantageous as significant amount of air might be trapped inside the soil-pile gap. Figure 11 shows the grout volume fraction at different time intervals for different inlet configurations under 500 kPa grout pressure and water-cement ratio of 0.4. Figure 11 shows the effect of a single inlet in each of the configuration shown with dotted lines in Fig. 9. Figure 11 shows the influence zone of individual inlets. Figure 11 (a), (b) and (c) shows 180° of the two-inlet configuration, 120° of the three-inlet configuration and 90° of the four-inlet configuration, respectively, with inlet at the center. It can be observed in Fig. 11 (b) and (c) that for three- and four-inlet configurations, there are large volumes of entrapped air in the soil-pile gap at the edges. This entrapped air could lead to reduction in grouting area and will lower grouting effectiveness. For the two-inlet configuration, even though the time taken for the grout to reach the same height of grout flow is higher compared to three- and four-inlet configurations (Fig. 10), the entrapped air volume is absent (see Fig. 11). The lack of entrapped air in the two-inlet configuration leads to the complete area between the soil-pile gap to be grouted, resulting in better grout strengths compared to the three- and four-inlet configurations. The increase in grouted area leads to an improvement in pile performance; hence, in practice, two-inlet configuration is used.

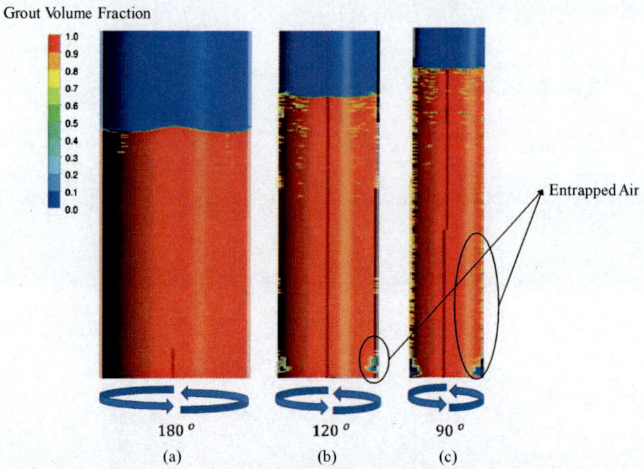

Fig. 11. Grout volume fraction in the pile–soil gap after 25 s: (a) two-inlet; (b) three-inlet; (c) four-inlet configurations (not to scale)

4 Conclusions

In this paper, a numerical analysis of grout flow in a narrow 10-mm soil-pile gap was conducted using the ANSYS Fluent software. Four different pile diameters of 800, 1000, 1200 and 1400 mm were considered and for each pile diameter, grouts with two different water-cement ratios of 0.4 and 0.5 were pumped into the soil-pile gap. Numerical results indicate that: 1) for the same pile diameter and grout pressure, the height of grout flow increases as the water-cement ratio increases from 0.4 to 0.5; 2) under the same grout pressure, the height of grout flow decreases as the pile diameter increases; 3) for a give pile diameter and water-cement ratio, the final grout height is mainly dependent on the grout pressure but insensitive to the inlet configuration; and 4) the two-inlet configuration is more effective in eliminating entrapped air at the bottom of the pile than the three- and four-inlet configurations.

Acknowledgment. The support from National Nature Science Foundation of China [NSFC (51608139 and 51678171)], the Peal River S&T Nova Program of Guangzhou (201806010095), the Science and Technology (S&T) Program of Guangzhou (201604016021), and the Department of Science and Technology of Guangdong Province (2015B020238014) is gratefully acknowledged.

References

Aoki, I.: Various construction methods and construction machines of embedded piles. [In Japanese.] Found. Eng. Equip. Mon. 26(2), 24–30 (1998).

Atzeni, C., Massidda, L., Sanna, U.: Comparison between rheological models for portland cement pastes. Cement Concrete Res. **15**, 511–519 (1985)

Bradshaw, A.S., and Baxter, C.D. : Design and construction of driven pile foundations: Lessons learned on the central artery/tunnel project. FHWA-HRT-05–159. Federal Highway Administration, McLean, VA, (2006)

Chan, W., Y., K., Jacobs, P., A and Mee, D., J. : Suitability of the k-ω turbulence model for scramjet flow field simulations. Int. J. Numer. Meth. Fluids **70**, 493–514 (2012)

Chen, H., Hu, H., Tang, M., Yang, X., Zhu, J.: Hybrid bored prestressed concrete cased piles: equipment and construction procedures. J. Constr. Eng. Manag. **144**(12), 06018006 (2018)

Cyr, M., Legrand, C., Mouret, M.: Study of the shear thickening effect of superplasticizers on the rheological behavior of cement paste containing or not mineral admixtures. Cement Concrete Res. **30**, 1477–1483 (2000)

ockray, T.J.: Thesis: Pile design and construction. Univ. of Queensland, Brisbane, Australia (2001)

Eriksson, M., Stille, H., Anderson, J.: Numerical calculations for prediction of grout spread with account for filtration and varying aperture. Tunneling and Underground Space Technol. **15**, 353–364 (2000)

FLUENT Manual, 2012. ANSYS/FLUENT Release Version 14.5 Ansys Inc., Documentation.

Kays, W., M., Crawford, M., E. and Weigand, B. Convective heat and mass transfer. fourth ed., McGraw Hill (2005).

Kishi, H., Kakue, T., Komine, H. and Murakami, S. Evaluation of Turbulent Flow conditions in rock cracks and Feasibility of a new grouting method for laminar and turbulent condition. 8th Asian Rock Mechanics Symposium, pp. 2661–2670 (2014).

CFD simulation of boiling flows using the volume-of-fluid method within OpenFOAM. Numerical Heat Transfer. Part A: Applications. 56, 631–646 (2005).

Lachemi, K., M., A., Hossain, V., Lambros, P., C., Nkinamubanzi, N. and Bouzoubaa, : Performance of new viscosity modifying admixtures in enhancing the rheological properties of cement paste. Cement Concrete Research. **34**(2), 185–193 (2004)

de Larrard, F., Ferraris, C., Sedran, F., T. : Fresh concrete: a Herschel – Bulkley material. Mater. Struct. **31**, 494 (1998)

Liovic, P., Liow, J., L. and Rudman, M. : A volume of fluid method for the simulation of metallurgical flows. ISIJ Int. **41**, 225–233 (2000)

Menter, F. R., Kuntz, M. and Langtry, R. (2003). "Ten years of industrial experience with the SST turbulence model". 4th International Symposium on Turbulence Heat and Mass Transfer. 625–632.

Moody, L.F.: Friction factors for pipe flow. Trans. ASME **66**, 671 (1944)

Nguyen, V., H., Remond, S., Gallias, J., L., Bigas, J., P. and Muller, P. : Flow of Herschel-Bulkley fluids through the Marsh cone. J. Non-Newtonian Fluid Mech. **139**, 128–134 (2006)

Nie, R., Leng, W., Yang, Q., Chen, Y.F.: An improved instrumentation method for PHC piles. Proc. Inst. Civ. Eng. Geotech. Eng. **169**(6), 494–508 (2016). https://doi.org/10.1680/jgeen.15.00138

Pillod, J., E. and Puckett, E., G. : Second-order accurate volume of fluid algorithms for tracking material interfaces. J. Comput. Phys. **199**, 465–502 (2004)

Renardy, M., Renardy, Y., Li, J.: Numerical simulation of moving contact line problems using a volume-of-fluid method. J. Comput. Phys. **171**(1), 243–263 (2001)

Rosquoet, F.A., Alexis, A., Khelidj, A., Phelipot. : Experimental study of cement grout: rheological behavior and sedimentation. Cement Concrete Res. **33**, 713–722 (2003)

Schlichting, H., Gersten, K.: Boundary-Layer theory, 8th edn. Springer-Verlag, Berlin, Heidelberg (2000)

Tattersall, G., H. and Banfill, P., F., G. : The rheology of fresh Concrete. Pitman, Boston (1983)

Wilcox, D., C. : Turbulence modelling for CFD, 3rd edn. DCW Industries, Incorporated, California (2006)

Author Index

S. S. Kim et al. (Eds.): GeoChina 2021, SUCI, p. 113, 2021.
https://doi.org/10.1007/978-3-030-80152-6